MATHEMATICAL FOUNDATION OF GIS

TOPICS IN ADVANCED GEOINFORMATICS

Series Editors: Jianya Gong *(Wuhan University, China)*
Huayi Wu *(Wuhan University, China)*

This book series aims to comprehensively cover the theory, methodology and applications of geospatial information science (GIScience), which mainly includes advanced topics on geographic information system (GIS), remote sensing (RS), and global navigation satellite system (GNSS).

The books in this series will include monographs and graduate level textbooks aimed at young researchers and postgraduates specialising in remote sensing, surveying and mapping, geography, urban planning and environmental science, global climate change and related majors.

Published

Vol. 2   *Mathematical Foundation of GIS*
by Wolfgang Kainz (*University of Vienna, Austria*) and
Huayi Wu (*Wuhan University, China*)

Vol. 1   *Urban Remote Sensing*
by Zhenfeng Shao (*Wuhan University, China*)

Topics in
Advanced
Geoinformatics
Volume 2

MATHEMATICAL FOUNDATION OF GIS

Wolfgang Kainz
University of Vienna, Austria

Huayi Wu
Wuhan University, China

高等教育出版社
HIGHER EDUCATION PRESS

World Scientific

Published by

Higher Education Press Limited Company
4 Dewai Dajie, Xicheng District
Beijing 100120, P. R. China

and

World Scientific Publishing Co. Pte. Ltd.
5 Toh Tuck Link, Singapore 596224
USA office: 27 Warren Street, Suite 401-402, Hackensack, NJ 07601
UK office: 57 Shelton Street, Covent Garden, London WC2H 9HE

Library of Congress Cataloging-in-Publication Data
Names: Kainz, Wolfgang, author. | Wu, Huayi, author.
Title: Mathematical foundation of GIS / Wolfgang Kainz, Huayi Wu.
Description: Singapore ; Hackensack, NJ : World Scientific, 2024. |
    Series: Topics in advanced geoinformatics ; Volume 2.
Identifiers: LCCN 2024017602 | ISBN 9789811292873 (hardcover) |
    ISBN 9789811292880 (ebook) | ISBN 9789811292897 (ebook other)
Subjects: LCSH: Geographic information systems--Mathematics. | Geomatics.
Classification: LCC G70.23 .K35 2024 | DDC 910.285--dc23/eng/20240514
LC record available at https://lccn.loc.gov/2024017602

British Library Cataloguing-in-Publication Data
A catalogue record for this book is available from the British Library.

For any available supplementary material, please visit
https://www.worldscientific.com/worldscibooks/10.1142/13829#t=suppl

Desk Editors: Balasubramanian Shanmugam/Steven Patt

Preface

Mathematics is an activity that has been performed by humans since thousands of years. The understanding of what mathematics is has changed over the centuries. In the beginning, mathematics was mainly devoted to practical calculations related to trade and land surveying. Over the centuries, mathematics has become a scientific discipline with many applications in all domains of life. *Geographic information system* (GIS) was introduced in the 1960s. It has become a major tool and science for the handling of spatial data, and has penetrated all walks of our daily lives. In this book, we will use the term geographic information system synonymously with spatial information system.

This preface gives a brief history of mathematics and explains how the different theories and branches of mathematics are rooted in logic and set theory. It also explains how the different chapters of this book can be read and used for a better understanding of the mathematical foundation of GIS.

The first known cultures that actively performed mathematical calculations in ancient history were the Sumerians, the Babylonians, the Egyptians, and the Chinese. In the beginning, mathematics was always related to practical problems of commerce, trading and surveying. This is the reason why the ancient cultures mainly developed practical solutions for arithmetic and geometric problems.

In the 5th century BC, the ancient Greeks started to do mathematics for its own sake, and to focus the scientific attention to mathematics as a science. The concepts of axioms and logical deduction were developed then. The first great example of this approach is *The Elements* of Euclid, the first textbook on geometry, which kept its validity until the 19th century when so-called non-Euclidean geometry was discovered.

The Indians and the Arabs further developed the number concept and trigonometry. In the 17th and 18th century, the concepts of calculus and analytical geometry were developed as a consequence of the intensive studies in physics and other natural sciences.

In the 19th century, mathematicians began to establish an axiomatic foundation of mathematical theories. Starting from a minimal set of axioms, statements (theorems) can be derived whose validity can be formally established by a proof. This axiomatic approach has been applied since then to formalize mathematics. Logic and set theory play an important role as the language of mathematics and the foundation principle of mathematical theories, respectively. The introduction of computer technology in the middle of the 20th century further stimulated the development of mathematics towards discrete mathematics and applications of mathematical logic in programming languages and database theory.

Logic is a formal language in which mathematical statements are written. It defines rules how to derive new statements from existing ones, and provides methods to prove their validity. *Set theory* deals with sets, the fundamental building blocks of mathematical structures, and the operations defined on them. The notation of set theory is the basic tool to describe structures and operations in mathematical disciplines. *Relations* define relationships among elements of sets. These relationships allow for instance the classification of elements into equivalence classes or the comparison of elements with regard to certain attributes. *Functions* (or mappings) are a special kind of relations and define how elements of a domain are mapped to their images (values) in a co-domain.

Sets whose elements are in certain relationships to each other or follow certain operations are mathematical *structures*. We distinguish between three major structures in mathematics, *algebraic, order,* and *topologic* structures. In sets with an algebraic structure we can do arithmetic (calculations), sets with an order structure allow the comparison of elements, and sets with a topologic structure allow to introduce concepts of convergence and continuity. Calculus, for instance, is based on topology. Often, sets carry more than one structure. The real numbers, for instance, carry an algebraic structure, an order structure, and a topologic structure. Results from algebraic topology (for example, simplicial and cell complexes, etc.) are used in the theory of GIS to structure spatial features. Figure 1 shows the sub-disciplines and their position in a general concept of mathematics.

On the top of the different structures and mixed structures, we find mathematical disciplines such as calculus, (analytic) geometry, probability theory and statistics. The classical theories that are of great importance in spatial data handling are (analytic) geometry, linear algebra, and calculus. With the introduction of digital technologies of GIS other branches of

Calculus	Probability theory	Statistics	Analytic geometry

Algebra	Ordered Sets	Topology

⇑ ⇑ ⇑

Algebraic structures	Order structures	Topologic structures
Structures		

⇑

Functions Relations

⇑

Set theory

⇑

Logic

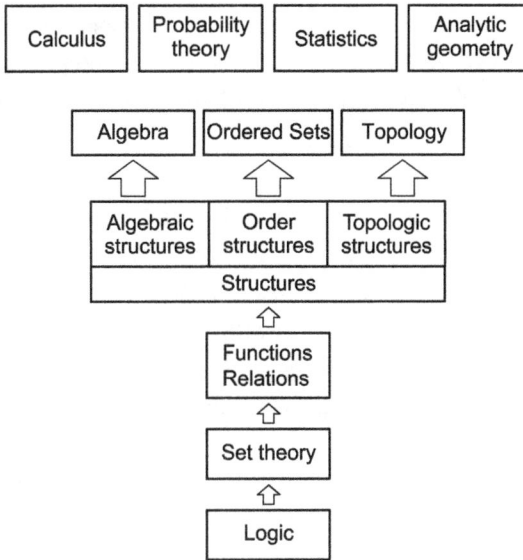

Figure 1 Sub-disciplines of mathematics and their relationships

mathematics became equally important, such as topology, graph theory, and the investigation of non-continuous discrete sets and their operations. The latter two fall under the domain that is usually called *finite mathematics* or *discrete mathematics* playing an important role in computer science and its applications.

The purpose of this book is to provide the reader with the mathematical knowledge needed when they have to deal with spatial data and GIS. Readers are expected to have a general knowledge of high school mathematics. The use of computers and software for the handling and processing of spatial data requires contents such as discrete mathematics and topology. The book is structured into 14 chapters.

Chapter 1 presents some philosophical considerations about space and time, the underlying concept of every spatial data handling in GIS. It shows that spatial modeling is built on solid mathematics as well as that there are challenging and interesting philosophical questions as to how to represent models of spatial features.

Chapters 2 to 4 deal with mathematical logic, the language and foundation of mathematics. Propositional and predicate logics are presented as well as logical inference, the methods of drawing logical conclusions from given facts.

Chapters 5 and 6 are introductions into the basic notions of sets, set operations, relations, and functions. These two chapters together with the three chapters on logic represent the foundation for the subsequent chapters dealing with mathematical structures.

Chapter 7 on coordinate systems and transformations builds the bridge between the foundation and the more advanced chapters on mathematical structures. Much of this chapter would normally be considered to belong either to (analytic) geometry or to linear algebra.

Chapters 8 to 11 present the highly relevant subjects of algebra, topology, ordered sets, and graph theory. These chapters address the mathematical core of many GIS functions from data storage and topological consistency to spatial analysis.

Uncertainty plays an increasingly important role in GIS. Chapter 12 addresses fuzzy logic and its applications in GIS. It shows how vague concepts can be formalized in mathematical language and how they are applied to spatial decision making.

Chapters 13 and 14 deal with probability theory and statistical discriminant analysis.

The book can be read in several ways as illustrated in Figure 2.

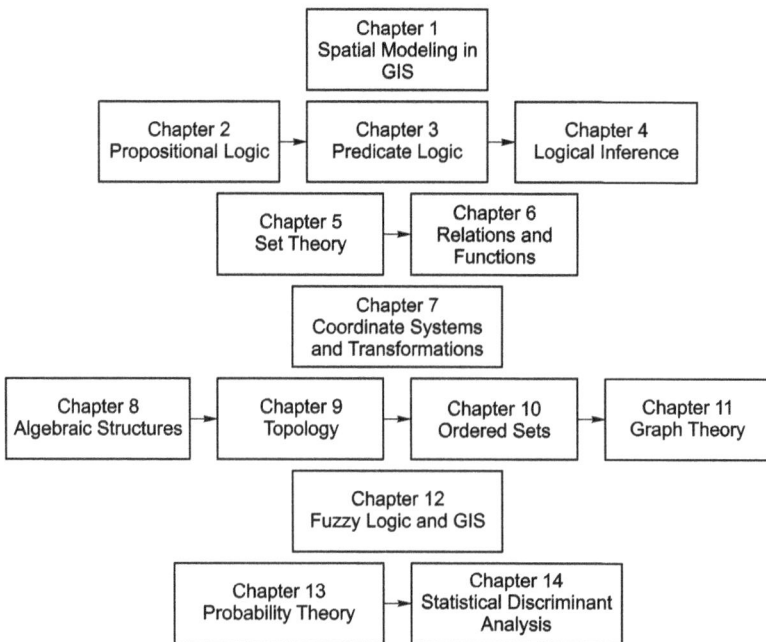

Figure 2 Different ways of reading this book

All readers should read Chapter 1. Readers interested in the logical and set theoretic foundations might want to read Chapters 2 to 4 and Chapters 5 and 6, respectively. For someone with a particular interest in more advanced structures, Chapters 8 to 11 will be of interest.

Chapters 7 and 12 can be read individually without losing too much of the context. Chapters 13 and 14 should be read by readers interested in probability theory and mathematical statistics. Both are important in remote sensing and machine learning environments. The best way, of course, is to read all the text from Chapters 1 to 14.

Wolfgang Kainz, *Vienna*
Huayi Wu, *Wuhan*
March 2024

About the Authors

Wolfgang Kainz is an emeritus professor of Cartography and Geographic Information Science, University of Vienna, and a visiting professor of Wuhan University. He received the PhD in Geographic Information Science at Graz Technical University. He has been engaged in geographic information science education for over 40 years, and his research focuses on spatiotemporal modeling, uncertainty, and topology, etc. He used to work in the International Cartographic Association and the International Society for Photogrammetry and Remote Sensing (ISPRS), and now serves as the chairman of the Austrian Cartographic Committee. He is the editor in chief of *ISPRS International Journal of Geo-Information*. In 2020, he won the Chinese Government Friendship Award and the Wang Zhizhuo Award from the International Society of Photogrammetry and Remote Sensing.

Huayi Wu is the Vice Director and the Chair Professor of the State Key Laboratory of Information Engineering in Surveying, Mapping and Remote Sensing, Wuhan University. He got his PhD in Photogrammetry and Remote Sensing in 1999 from Wuhan University and his thesis was awarded one of the 100 Excellent Doctoral Theses by the Ministry of Education in 2002. His research interests include high-performance geospatial computing, spatiotemporal data mining and visualization. He has served as the principal investigator for many research projects supported by the Ministry of Science and Technology, Ministry of Education, and National Natural Science Foundation of China, etc. He is the associate editor of *ISPRS International Journal of Geo-Information*, and the editorial board member of *Geo-spatial Information Science*, and *Transaction in GIS*.

Contents

Chapter 1

Spatial Modeling in GIS

We live in a constantly changing world. What we perceive with our senses are processes, states and events that happen or exist in the real world. They may be natural or human-made, and are called real-world phenomena. In our brains, we process the input from our senses, which leads to mental models, learning, cognition, and knowledge. Everything that happens in the real world and that we process through our senses leads to models of reality that we create for ourselves. Usually, we agree on common principles and interpretations derived from experience, research or learning that enable us to reach a consensus on the perception of real-world phenomena.

In an analog way as we perceive and create models of the real world in our minds, we design and create models of subsets of reality in spatial databases. This chapter deals with spatial modeling, a process that maps subsets of the real world to abstract models. We discuss fundamental principles of space and time and their bearing on geographic information systems.

1.1 Real World Phenomena and Their Abstractions

In the real world, we distinguish between natural and human-made *phenomena*. Natural phenomena exist independently from human actions and are subject only to the laws of nature. Examples are the landscape (topography), the weather, or the natural processes that shape and influence them. Human-made phenomena are objects that were created by human activities through construction or building processes.

Based on these phenomena we develop high-level abstract models for particular purposes and applications. *Features* (abstractions of phenomena) populate these models that are usually organized in *layers*. Examples of such models are a cadaster, topography, soil, hydrography, land cover, or land use.

The fact that these models are abstractions of the real world can be illustrated by the example of a cadaster. Let us assume that a cadaster is a legal and organizational framework for the management of land. It is a very important, clearly described and understood concept. Yet, we do not see a cadaster when we look around us. What we see are real world phenomena such as buildings, roads, fences enclosing pieces of land, and people. A cadaster abstracts from certain phenomena and their relationships to create something new that is real in a given context.

Layers are an ordering principle for phenomena. Again, when we look around us, we do not see the world in layers. Yet, they are used to organize phenomena of the real world in such a way that we classify them according to a perceived purpose or characteristic into subsets (layers) that allow us to deal with them efficiently.

1.1.1   Spatial Data and Information

Humans perceive signals (data) through their senses, process them and extract information that leads to knowledge and wisdom. In order to conceptualize mental models of the real world, we need to categorize the phenomena we observe. These phenomena exist in space and time and have, therefore, a spatial (geometric) and temporal extent. They possess certain thematic characteristics (or attributes) that allow us not only to extract spatiotemporal information, but also thematic information. The thematic information of real-world phenomena that we extract by mental processes is the basis for the definition of layers.

In GIS, we focus on spatial information extracted from spatial data by data processing and spatial analysis. We need to store representations of the phenomena in a database. To achieve this, we need to define features, organize them into layers, collect (spatial and attribute) data about them and enter the data into a database. Data processing and spatial analysis extract information from the data. The following summarizes these.

- Real-world phenomena have a spatiotemporal extent and possess thematic characteristics (attributes).

- A (spatial) feature is a representation of a real-world phenomenon.
- Spatial data are computer representations of spatial features.
- Spatial data handling extracts (spatial) information from spatial data.

1.2 Concepts of Space and Time

Space and time are two closely related concepts that have been the subject of philosophical and scientific considerations since the dawn of humankind. The space that humans live in is the three-dimensional (Euclidean) space as a frame of reference for our senses of touch and sight. Of all possible physical and mathematical spaces, this is the unique space that is illustrative and that we perceive as being real. Time is a measure for change in our immediate experience. Usually, we assume time to be of a continuous linear nature extending from the past, through the present into the future.

Space and time (at least as we perceive them) are so well known and appear to be given beyond any doubt that we hardly ever contemplate their structures and characteristics in our everyday lives. When we deal with information systems that process and manipulate spatiotemporal data, we need clear and well-understood models of space and time. The following sections describe how concepts of space and time developed in Western philosophy and physics. We discuss them according to three epochs: (i) pre-Newtonian concepts, (ii) the Newtonian and classical concepts, and (iii) contemporary concepts of space and time.

1.2.1 Pre-Newtonian Concepts of Space and Time

The concepts of space and time of this epoch are mainly dominated by the ideas of Greek philosophers about the logical conditions for things to change and the structure of the world in which change occurs.

Heraclitus (544–483 BC) of Ephesus (western Turkey) studied the problem of change, i.e., how can the identity of things be preserved when they change? He stated that everything flows, nothing remains, and the only thing that really exists is change (processes). The quotes "Everything flows" and "We cannot step into the same river twice" are attributed to him.

At about the same time, Parmenides (late 6th or early 5th century BC) of Elea (southern Italy) developed a completely opposite philosophy of the non-existence of the void. He postulated (through deductive reasoning) that change does not exist and that the real world (the real being) is *plenum* (a solid complete compact being), immutable, without change, and eternal. A void (or empty space, i.e., something non-existent) does not exist. What we perceive as change is a delusion of our senses. Parmenides' ideas were further developed and "proven" by his student Zeno (490–425 BC) of Elea.

One of Zeno's famous "proofs" that change and movement cannot exist is known as the race between Achilles and the tortoise. Assuming that there are an infinite number of points on a straight line labeled as A, B, C, ... from the starting point to the endpoint in sequence. The tortoise gets a head-start and begins the race at point B, whereas Achilles starts from point A. When Achilles reaches point B, the tortoise has already moved on to point C. When Achilles reaches point C, the tortoise has moved on to another point, and so on. The lead of the tortoise gets smaller and smaller, *ad infinitum*. We get an infinite number of (smaller and smaller) leads.

The argument is now: In order to reach the tortoise, Achilles must catch up an infinite number of (finite in length) leads. It is impossible to run this infinite number of short distances, because Achilles would have to run infinitely far (or forever). Therefore, it is impossible for Achilles to catch the tortoise.[1] Since we can easily catch a tortoise, when we run against it, we end up with a paradox. This proves the assumption that movement and change are real, leads to contradictions. Therefore, movement and change are impossible.

Democritus (460–370 BC) did not accept the non-existence of change as postulated by Parmenides. Space is an absolute and empty entity existing independently from the atoms that fill the space. Atoms are indivisible real things; they are immutable and eternal, and have different sizes and weights. There is no empty space within atoms. An atom is a *plenum*. Objects are formed as a collection of atoms. The importance of the Atomist theory is evident in today's modern particle physics.

------

[1] The solution of the paradox lies in the fact that an infinite series can converge to a finite value, i.e., in our case the point where Achilles passes the tortoise. This mathematical result was, however, not known until the 17th century.

| Tetrahedron (fire) | Octahedron (air) | Dodecahedron | Icosahedron (water) | Cube (earth) |

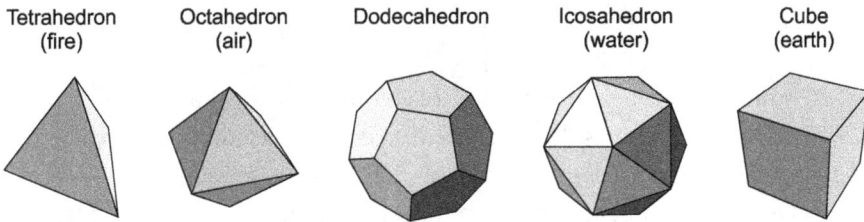

Figure 1.1   Platonic solids as building blocks of matter

Greek mathematics was strongly dominated by the Pythagorean number theoretic approach. It was essentially arithmetic based on (philosophical) properties of numbers, counting, and the ratios between numbers. The discovery of irrational numbers such as $\sqrt{2}$ (the length of the diagonal in the unit square) shook the foundation of Greek mathematics that was based on counting in natural units. The need for a truly geometrical description of the world became apparent.

The great philosopher Plato (427–347 BC) and one member of his school, Euclid (around 300 BC), laid the foundation for a new geometric modeling of the real world. This geometric model of matter is based on symmetric geometric solids known as the Platonic solids. Matter consists of four elements: earth, air, fire, and water. Each element is made of particles, i.e., solids (see Figure 1.1).

In his book, *The Elements*, Euclid developed a mathematical theory of geometry that remained valid until the late 19th century. Euclidean geometry was considered a true description of our physical world until it was discovered that many consistent geometric systems are possible, some of which are non-Euclidean, and that geometry is not a description of the world but yet another formal mathematical system with no necessary reference to real world phenomena.

1.2.2   Classical Concepts of Space and Time

The time between the classical Greek period and the rise of modern science was dominated by the philosophy and teaching of Aristotle (384–322 BC). According to his ideas, empty space is impossible, and time is the measure of motion with regard to what is earlier and later. Space is defined as the limit of the surrounding body towards what is surrounded.

Following from this approach space can be conceptualized in two possible ways:

- **Absolute space**: Space as a set of places. It is an absolute real entity, the container of all things. Its structure is fixed and invariable. Generally, this is considered the space as described by Euclidean geometry.
- **Relative space**: Space is a system of relations. It is the set of all material things, and relations are abstracted from them. Space is a property of things or things have spatial properties.

The rise of modern science took shape in the works of Nikolaus Copernicus (1473–1543) (heliocentric system stating that the sun is the center of our planetary system), Johannes Kepler (1572–1630) (mathematical foundation of the heliocentric system), and Galileo Galilei (1564–1642) (foundations of mechanics) in the 16th and 17th centuries.

Isaac Newton (1643–1727) was a brilliant scientist (dynamics theory) and philosopher. In his philosophy, he was an outspoken proponent of the concept of absolute space, although it strictly contradicts his dynamics theory. The concept of absolute space remained dominant until the late 19th century.

Gottfried Wilhelm Leibniz (1646–1716) on the other hand sustained the concept of relative space. For him, space is a system of relationships between things. It is interesting to note that both Newton and Leibniz are the founders of mathematical calculus.

One of the greatest philosophers, Immanuel Kant (1724–1804), claimed that space and time are not empirical physical objects or events. They are merely *a priori* true intuitions, not developed by experience, but used by us to relate and order observations of the real world. Space and time have *empirical reality* (they are absolute and *a priori* given) and *transcendental idealism* (they belong to our conceptions of things but are not part of the things). We cannot know anything about the things as such. In this regard, Kant can be seen as a proponent of absolute space, yet in a far more elaborate and sophisticated way than the previous philosophical approaches.

1.2.3   Contemporary Concepts of Space and Time

The development of modern physics (field theory, theory of relativity, quantum theory) and mathematics (non-Euclidean geometry) led to the conclusion that traditional Euclidean geometry (describing the three-dimensional space of our perception) is only an approximation to the real nature of the world.

The field theories proposed by Michael Faraday (1791–1867) and James Clerk Maxwell (1831–1879) led to the assumption that space is not empty, but is filled with energy. Therefore, a material existence of space is strongly supported by these theories.

As a consequence of the special and general theory of relativity by Albert Einstein (1879–1955), space and time cannot anymore be considered as two separate entities. We speak of space–time, which is considered a four-dimensional space that can only be described by non-Euclidean geometry. Quantum mechanics states the principle of uncertainty and the discrete character of matter and energy. It has become evident that the space of our perception is not necessarily identical with the microscopic (sub-atomic) space or the space of cosmic dimensions.

1.2.4   Concepts of Space and Time in Spatial Information Systems

Spatial information is always related to geographic space, i.e., large-scale space. This is the space beyond the human body, space that represents the surrounding geographic world. Within such space, we move around, we navigate in it, and we conceptualize it in different ways. Physical geographic space is the space of topographic, cadastral, and other features of the geographic world. Geographic information system technology is used to manipulate objects in geographic space, and to acquire knowledge from spatial facts.

The human understanding of space, influenced by language and cultural background, plays an important role in how we design and use tools for the processing of spatial data. In the same way as spatial information is always related to geographic space, it relates to the time whose effects we observe in the changing geographic world around us. We are less interested in pure philosophical or physical considerations about time or space–time, but more in the observable spatiotemporal effects that can be described, measured and stored in information systems.

1.3   The Real-World and Its Models

As mentioned in the previous sections, we always create models of the real world in our minds. When we want to acquire, store, analyze, visualize, and exchange information about the real world, we use other media and means than just interpersonal communication. We need representations of our mind models, i.e., models of the real world that can be used to acquire, store, analyze and transfer information about real world data.

The most common ones of these models are—in historic sequence— maps and databases. Both have distinct characteristics, advantages and disadvantages. Whereas maps usually have been used to picture real-world phenomena, databases can be used to represent real and virtual worlds.

Real worlds are subsets of the reality that we perceive. Virtual worlds are computer-generated "realities" that exist only as potentialities with no counterpart in the real world. Yet, we can visualize them, navigate through them and perceive them as "real" (therefore the term virtual reality). There is no difference between real and virtual world models with regard to their representation. The only difference is that the former is a model of something that exists in the real world and the latter is a model of something that exists only in a virtual (physically non-existent) world.

1.3.1   Maps

The best known (conventional) models of the real world are maps. Maps have been used for thousands of years to represent information about the real world. We know maps from ancient Mesopotamia, Egypt, and China, through the Roman times, the Medieval Ages until the present. In modern times, they are usually drawn on paper or other permanent materials and function as data storage and visualization medium. Their conception and design have developed into an art and science with a high degree of scientific sophistication and artistic craftsmanship. Maps have proven to be extremely useful models of reality for many applications in various domains.

Yet, maps are two-dimensional (flat) and static. It is not easy to visualize three-dimensional dynamic features without considerable abstractions in the spatial and temporal domains. We distinguish between topographic and thematic maps. Other cartographic products that are not maps are often used to represent three-dimensional and dynamic phenomena. Such products are, for instance, block diagrams, animations, and panoramic views.

A disadvantage of maps is that they are restricted to two-dimensional static representations, and that they are always displayed in a given scale. The map scale determines the spatial resolution of the graphic feature representation. The smaller the scale, the less detail a map can show. The accuracy of the primary data, on the other hand, puts limits to the scale in which a map can be drawn. The selection of a proper map scale is one of the first and most important steps in a map conception.

A map is always a graphic representation at a certain level of detail, which is determined by the scale. The process to derive less detailed representations from a detailed one is called map generalization (or cartographic generalization). Map sheets have physical boundaries, and features spanning different map sheets have to be cut into pieces.

The use of computers in map making is an integral part of modern cartography. The role of the map changed accordingly. Increasingly, maps lost their role as data storage. This role is taken over by databases. What remains is the visualization function of maps.

1.3.2   Databases

Spatial databases store representations of spatial phenomena in the real world to be used in a geographic information system. They are also called GIS databases or geodatabases. In the design of a database, we distinguish between different levels of definition. A complete database definition at a particular level is called a *database schema*.

The assumption for the design of a spatial database schema is that spatial phenomena exist in a two- or three-dimensional Euclidean space. All phenomena have various relationships among each other and possess spatial (geometric), thematic and temporal attributes. Phenomena are classified into thematic layers depending on the purpose of the database. This is usually described by a qualification of the database, for example, cadastral, topographic, land use, or soil database.

The representations of spatial phenomena (i.e., spatial features) are stored in a *scaleless* and *seamless* manner. Scaleless means that all coordinates are world coordinates given in units that are normally used to reference features in the real world (geographic coordinates, meters, feet). From there, calculations can be easily performed and any (useful) scale can be chosen for visualization.

It must be noted, however, that scale plays a role when data are captured from maps as data source. Here, the scale of the source map determines the accuracy of the feature coordinates in the database. Likewise, the accuracy of measurements in field surveys determines the quality of the data. A seamless database does not show map sheet boundaries or other partitions of the geographic space other than imposed by the spatial features themselves.

It is easy to query a database, and to combine data from different layers (spatial join or overlay). Spatiotemporal databases consider not only the spatial and thematic but also the temporal extent of the features they represent. Various spatial, temporal and spatiotemporal data models have been developed.

1.3.3   Space and Time in Real-World Models

In modern physics, it is common to speak of space–time to express the close connection that exists between space and time according to the special and general theory of relativity. Here, we do not consider the physical characteristics of space and time, but focus on the (simplified) ways of representing spatial phenomena in a GIS database.

In general, modeling can be described as creating a structure preserving mapping (morphism) from a domain to a co-domain. In our case, the domain is the real world, and the co-domain is a real-world model. Such a mapping normally creates a "smaller" (i.e., abstracted, generalized) image of the original. Structure preserving means that the elements of the co-domain (spatial features) behave in the same (however simplified or abstracted) way as the elements of the domain (spatial phenomena). Figure 1.2 illustrates the principle of spatial modeling.

As mentioned above, we consider space to be the three-dimensional Euclidean space of our common sense. All phenomena exist in this space and undergo changes, which we perceive as the passing of time. In this sense, time is modeled indirectly as changes in (spatial or thematic) attributes of the features.

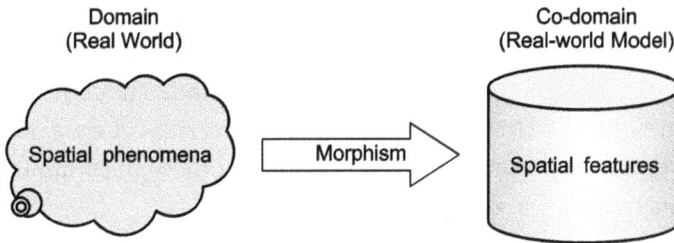

Figure 1.2   Spatial modeling is a structure preserving mapping from the real world to a spatial model

We call a spatial data model that also considers time a spatiotemporal data model. Sometimes such a model is addressed as four-dimensional model, giving the impression that time is the fourth dimension in addition to the three spatial dimensions. It is, however, better to call it spatiotemporal instead of four-dimensional. Firstly, time is not of the same type as the spatial dimensions; it has a distinctly different quality. Secondly, the term "four-dimensional" only makes sense when we always would consider three spatial dimensions. In most of the cases, however, the spatial data model only considers two dimensions.

1.4   Real-World Models and Their Representation

A spatial database holds a digital representation of the real world. The feature space for a spatial database is a geometric space in which we model features at various levels of detail. A good data model should allow for multiple representations of spatial features. The transformation of a representation from a detailed to a less detailed version is called model generalization. This is different from cartographic generalization where a graphic representation (digital or analog) at a smaller scale is derived from a large-scale data set or map under certain merely graphic constraints.

Database creation is a two-step process. Firstly, the database is designed by defining a database schema that identifies types but no occurrences. Secondly, the database is filled with real data. We are particularly interested in properties of the geometric space that remain invariant under certain transformations. This is important to guarantee a consistent representation of the features in a database. These properties are related to topology.

Some features possess a degree of uncertainty in either their thematic or spatial extent. For example, soil types do not have crisp boundaries. Linguistic expressions of vagueness or uncertainty such as "moderate slope" or "close to Vienna" are often a better analysis approach than clear cut class boundaries. We have to devise proper means to take care of these uncertainties in our spatial data models. Fuzzy logic provides the tools.

Spatial data exist not only in space but also in time. They carry temporal characteristics that data models must be able to handle. Several spatiotemporal data models have been proposed.

1.4.1 Spatial Data Models

We distinguish two major types of spatial data models, field- and object-based models. *Field-based models* consider spatial phenomena to be of a continuous nature where in every point in space a value of the field can be determined. Examples of such phenomena are temperature, barometric pressure, or elevation. *Object-based models* consider space to be populated by well distinguishable, discrete, bounded objects with the space between objects potentially being empty. Examples are a cadaster with clearly identifiable objects like parcels and buildings.

Field versus object can be viewed as a manifestation of the philosophical conception of plenum versus atomic space (see Section 1.2.1), or as in modern physics, *wave* versus *particle* (see Section 1.2.3). In GIS, fields are usually implemented with a tessellation approach (usually raster or grid), while objects with a (topological) vector approach. The following sections briefly illustrate the two different model types.

1.4.1.1 *Field-based Models*

The underlying space for a field-based model is usually taken as the two- or three-dimensional Euclidean space. A field is a computable function from a geometrically bounded set of positions (in 2D or 3D) to some attribute domain. Computable means that for every position within the geometric bounds a value can be determined by either measurement or computation. A field-based model consists of a finite collection of such fields.

Fields can be discrete, continuous and differentiable. Discrete fields represent features with boundaries; continuous fields are used for features where the underlying function is continuous, such as temperature, barometric pressure or elevation. If the field function is differentiable, we can for instance compute the slope at every position of an elevation field.

Though geometrically bounded, the domain of a field is still an infinite set of positions. Computers cannot represent field values for an infinite number of positions. Therefore, we must accept an approximation. The standard way to obtain this is to finitely represent the geometrically bounded space through a subdivision into a regular or irregular tessellation, consisting of either square (or cubic) or triangular (tetrahedral) parts. These individual parts are called *locations*. Representations of locations in a GIS are raster or grid cells (pixels or voxels).

To represent spatial features using a field-based approach we have to perform the following steps:

(1) Define or use a suitable model for the underlying space (tessellation).
(2) Find suitable domains for the attributes.
(3) Sample the phenomena at the locations of the tessellation to construct the spatial field functions.
(4) Perform analysis, i.e., compute with the spatial field functions.

1.4.1.2   Object-based Models

Object-based models decompose the underlying space into identifiable, describable objects that have spatial, thematic, and temporal attributes, as well as relationships among each other. The space outside the objects is empty. Examples of objects are buildings, cities, towns, districts or countries; attributes are, for instance, the number of floors, population, boundary or area.

In a GIS database implementation, objects are represented as a structured collection of geometric primitives (points, lines, polygons, and volumes) under geometric, thematic and topological constraints.

Object-based models are discrete models. Operations in the model always refer to the manipulation of individual objects or sets of objects. Manipulations concern the spatial, thematic, topological or temporal domain. Accordingly, they are realized through geometric, attribute manipulations or topological operations.

Topology plays a major role in object-based models. It is the "language" that allows us to specify and enforce consistency constraints for spatial databases. The majority of object-based models are two-dimensional. Recently, three-dimensional data models have been introduced. Their topology is more difficult to handle than in the standard two-dimensional cases.

1.4.2 Spatiotemporal Data Models

Beside geometric, thematic and topological properties, spatial data also possess temporal characteristics. It is, for instance, interesting to know who were the owners of a land parcel in 1980, or how did land use of a particular piece of land change over the last 20 years.

Spatiotemporal data models are data models that can also handle temporal information in spatial data. Several models have been proposed. The most important ones will be discussed briefly. Before we describe the major characteristics of the spatiotemporal data models, we need a framework to describe the nature of time itself. Time can be characterized according to the following properties:

- **Time density**: Time can be discrete or continuous. *Discrete time* is composed of discrete elements (seconds, minutes, hours, days, months, or years). In *continuous time*, for two points in time, there is always another point in between. We can also structure time by *events* (points in time) or *states* (time intervals). When we represent states by intervals bounded by nodes (events) we can derive temporal relationships between events and states such as "before", "overlap", "after", etc.
- **Dimensionality of time**: *Valid time* (or world time) is the time when an event really happened. *Transaction time* (or database time) is the time when the event was recorded in the database.
- **Time order**: Time can be *linear*, extending from the past to the present, and into the future. We know also *branching time* (possible scenarios from a certain point in time onwards) and cyclic time (repeating cycles such as seasons or days of a week).
- **Measures of time**: A *chronon* is the shortest non-decomposable unit of time that is supported by a database (e.g., a millisecond). The life span of an object is measured by a (finite) number of chronons. *Granularity* is the

precision of a time value in a database (e.g., year, month, day, second, etc.). Different applications require different granularity. In cadastral applications, time granularity can be a day; in geological mapping, time granularity is more likely in the order of thousands of years.

- **Time reference**: Time can be represented as *absolute* (fixed time) or *relative* (implied time). Absolute time marks points on the timeline where events happen (e.g., "6 July 1999 at 11:15 p.m."). Relative time is indicated relative to other points in time (e.g., "yesterday", "last year", "tomorrow", which are all relative to "now", or "two weeks later", which may be relative to an arbitrary point in time).

In spatiotemporal data models, we consider changes of spatial and thematic attributes over time. In data analysis, we can keep the spatial domain fixed and look only at the attribute changes over time for a given location in space. We would, for instance, be interested in how land cover changed for a given location or how the land use changed for a given land parcel over time, provided its boundary did not change.

On the other hand, we can keep the attribute domain fixed and consider the spatial changes over time for a given thematic attribute. In this case, we could be interested to see which locations were covered by forest over a given period.

Finally, we can assume both the spatial and attribute domain variable and consider how an object changed over time. This actually leads to notions of object motion, and these are a subject of research, with for instance applications in traffic control, mobile telephony, wildlife tracking, disease control, and weather forecasting. Here, the problem of *object identity* becomes apparent. When does a change or movement cause an object to disappear and become a new one? In the following, we describe the major characteristics of some popular spatiotemporal models.

1.4.2.1 Space–Time Cube Model

This model is based on a two-dimensional space (spanned by the x- and y-axis) whose features are traced through time (along the z-axis) thereby creating a space–time cube. The traces of objects through time create a worm-like trajectory in the space–time cube. This model potentially allows absolute, continuous, linear, branching and cyclic time. It supports only valid time. The attribute domain is kept fixed and the spatial domain is variable.

1.4.2.2   *Snapshot Model*

In the snapshot model, layers of the same theme are time-stamped. For every point in time that we would like to consider, we have to store a layer and assign the time to it as an attribute. We do not have any information about the events that caused different states between layers. This model is based on a linear, absolute, and discrete time. It supports only valid time and multiple granularity. The spatial domain is fixed (field-based) and the attribute domain is variable.

1.4.2.3   *Space–Time Composite Model*

The space–time composite model starts with a two-dimensional situation (plane or layer) at a given start time. Every change of features that happens later is projected onto the initial plane and intersected with the existing features, thereby creating an incrementally built polygon mesh. Every polygon in this mesh has its attribute history stored with it. The space–time composite model is based on linear, discrete, and relative time. It supports both valid and transaction time, and multiple granularity. It keeps the attribute domain fixed and the spatial domain variable.

1.4.2.4   *Event-based Model*

In an event-based model, we start with an initial state and record events (changes) along the time line. Whenever a change occurs, an entry is recorded. This is a time-based model. The spatial and thematic attribute domains are secondary. The model is based on discrete, linear, and relative time. It supports only valid time and multiple granularity.

1.4.2.5   *Spatiotemporal Object Model*

This model is based on spatio-temporal objects that are a complex of spatio-temporal atoms. Both objects and atoms have a spatial and a temporal extent. The model is based on discrete, absolute, and linear time. It supports valid and transaction time as well as multiple granularity.

1.5 Summary

Geographic information systems process spatial data to derive information from these data in a database. To work with these systems, we need models of spatial data as a framework for database design. These models address the spatial, thematic and temporal dimensions of real-world phenomena. An understanding of the principle concepts of space and time rooted in philosophy, physics, and mathematics is a necessary prerequisite to develop and use spatial data models.

We know two major approaches to spatial data modeling, the analog map approach, and the spatial databases. Today, the function of maps as data storage (map as a database) is increasingly taken over by spatial databases. In databases, we store representations of phenomena of the real world. These representations are abstractions according to selected spatial data models. We know two fundamental approaches to spatial data modeling, field-based and object-based models. Both have their merits, advantages and disadvantages for particular applications.

Consistency is an important requirement for every model. Topology provides us with the mathematical tools to define and enforce consistency constraints for spatial databases, and to derive a formal framework for spatial relationships among spatial objects.

Spatial data not only possess spatial and thematic attributes, but also extend into the temporal domain. A model of time for spatial information is an important ingredient for any spatial data model, thus leading to what is called spatiotemporal data models.

Chapter 2

Propositional Logic

Propositional logic deals with assertions or statements that are either true or false and operators that are used to combine them. Such statements are called propositions. Any other statement for which we cannot establish whether it is true or false is not the subject of logic. This chapter explains the principles of propositional logic by introducing the concepts of proposition, propositional variable, propositional form, and logical operators. The translation of natural language into propositions and the establishment of their truth-values with the help of truth tables are shown as well.

2.1 Assertion and Proposition

Propositional logic deals with statements that are either true or false. Here, we will only deal with a two-valued logic. This is the logic on which most of the mathematical disciplines are built, and which is used in computing (a bit can only assume two states, on or off, one or zero).

Definition 2.1 (Assertion and Proposition). An *assertion* is a statement. If an assertion is either true or false, but not both,[1] we call it a *proposition*. If a proposition is true, it has a truth-value of true; if it is false, it has a truth-value of false. Truth-values are usually written as true, false, or T, F, or 1, 0. In the following sections, we will use the 1, 0 notation for truth-values.

---

[1]We call a logic in which assertions are either true or false a *two-valued logic*. The *law of the excluded middle* characterizes a two-valued logic.

Example 2.1 The following statements illustrate the concept of assertion, proposition and truth-values. The following assertions are propositions:
(1) "It rains".
(2) "I pass the exam".
(3) "$3 + 4 = 8$".
(4) "3 is an odd number".
(5) "7 is a prime number[2]".

Assertions (1) and (2) can be true or false. Proposition (3) is false, and (4) and (5) are true. The following statements are not propositions:
(6) "Are you at home"?
(7) "Use the elevator"!
(8) "$x + y < 12$".
(9) "$x = 6$".

(6) and (7) are not assertions (they are a question and a command, respectively), and therefore they cannot be propositions. (8) and (9) are assertions, but not propositions. Their truth-values depend on the values of the variables x and y. Only when we replace the variables with some values, the assertions become propositions.

Often, we have to be more general in expressing assertions. For this, we use propositional variables and propositional forms.

Definition 2.2 (Propositional Variable). A *propositional variable* is an arbitrary proposition whose truth-value is unspecified. We use upper case letters $P, Q, R, \ldots$ for propositional variables.

We can combine propositions and propositional variables to form new assertions by using words such as "and", "or", and "not".

Example 2.2 "Beer is good and water has no taste" is a combination of the two propositions "Beer is good" and "Water has no taste" using the connector "and". "P or not Q" is a combination of the propositional variables P and Q using the connectors "or" and "not".

---

[2] A *prime number* is any natural number n that can only be divided by 1 and n.

2.2 Logical Operators

In the example above, P and Q are called *operands*, and the words "and", "or", and "not" are *logical operators*, or *logical connectives*. An operator such as "not" that operates only on one operand is called *unary operator*; those that operate on two operands such as "and" and "or" are called *binary operators*.

Definition 2.3 (Propositional Form). A *propositional form* is an assertion that contains at least one propositional variable. We use upper case Greek letters to denote propositional forms, $\Phi(P, Q, \ldots)$.

When we substitute propositions for the propositional variables of a propositional form, we get a proposition. When we use logical connectives to derive new propositions from existing ones, the truth-values of the new propositions depend on the logical connective and the truth-values of the existing propositions.

Example 2.3 Let P stand for "Vienna is the capital of Austria" and Q stand for "Two is an odd number", then the propositional form in Example 2.2 "P or not Q" becomes the proposition "Vienna is the capital of Austria or two is not an odd number".

Logical operators are used to combine propositions or propositional variables. Table 2.1 shows the most common operators.

To determine the truth-value of a combined proposition we need to look at every possible combination of truth-values for the operands. This is done using *truth tables* that are defined for every operand. Table 2.2 shows the truth

Table 2.1 Logical operators

Logical operator	Symbol	Read or written as
Conjunction	$\wedge$	and
Disjunction	$\vee$	or
Exclusive or	$\oplus$	either … or but not both
Negation	$\neg$	not
Implication	$\Rightarrow$	implies, if … then …
Equivalence	$\Leftrightarrow$	equivalent, … if and only if …, iff[*]

Note: [*]The term "iff" short for "if and only if" is used only in written mathematical texts.

Table 2.2 Truth tables for logical operators

Conjunction			Disjunction			Exclusive or			Negation		Implication			Equivalence		
P	Q	$P \wedge Q$	P	Q	$P \vee Q$	P	Q	$P \oplus Q$	P	$\neg P$	P	Q	$P \Rightarrow Q$	P	Q	$P \Leftrightarrow Q$
0	0	0	0	0	0	0	0	0	0	1	0	0	1	0	0	1
0	1	0	0	1	1	0	1	1	1	0	0	1	1	0	1	0
1	0	0	1	0	1	1	0	1			1	0	0	1	0	0
1	1	1	1	1	1	1	1	0			1	1	1	1	1	1

tables for the most common logical operators. We use the symbols "0" for false and "1" for true.

Negation is a unary operator, i.e., it applies to one operand, and changes the truth-value of a proposition. The other operators apply to two operands. The conjunction (or logical and) is only true if both operands are true. The disjunction (or inclusive or) is true whenever at least one of the operands is true. The exclusive or is only true if either one or the other operand is true, but not both.

When we use the English term "or" we do not make explicit whether we mean the inclusive or exclusive or. It usually follows from the context what we mean. In mathematics, we cannot operate in this way. Therefore, we must make a distinction between inclusive or and exclusive or.

In the statement "I go to work or I am tired" the operator indicates an inclusive or. I can go to work and I can be tired at the same time. However, when we say that "I am alive or I am dead" we clearly mean an exclusive or. A person cannot be alive and dead at the same time.[3]

In the implication $P \Rightarrow Q$ we call P the *premise*, *hypothesis*, or *antecedent*, and Q the *conclusion* or *consequence*. The implication can be read in many different ways:
"If P, then Q".
"P implies Q".
"P is a sufficient condition for Q".
"Q if P".
"Q follows from P".
"Q provided P".

---

[3]We exclude here the possibility of being a zombie, a state of existence (the living dead) that appears frequently in horror movies.

"Q is a logical consequence of P".
"Q whenever P".

If $P \Rightarrow Q$ is an implication then $Q \Rightarrow P$ is called the *converse* and $\neg Q \Rightarrow \neg P$ is called the *contrapositive*.

Example 2.4   Let us consider the implication "If it rains, then I get wet". The converse of this implication reads as "If I get wet, then it rains", and the contrapositive is "If I do not get wet, then it does not rain".

In natural language, the implication expresses a causal or inherent relationship between a premise and a conclusion. The statement "If I take a shower, then I will get wet" clearly states a causal relationship between taking a shower and getting wet. The statement "If this is an airplane, then it has wings" expresses a property of airplanes.

In propositional logic, there need not be any relationship between the premise and the conclusion of an implication. We have to keep this in mind in order not to get confused by some propositions.

Example 2.5   Let us take P as "the moon is larger than the earth" and Q as "the sun is hot". The implication "If the moon is larger than the earth, then the sun is hot" is true, although there is no relationship whatsoever between the two propositions. The implication is true because P is false and Q is true. According to the truth table for implications, anything (either a true or a false statement) can follow from a false proposition.

Two propositions that have the same truth-values are said to be logically equivalent, denoted by $P \Leftrightarrow Q$. $P \Leftrightarrow Q$ can be read in different ways:
"P is equivalent to Q".
"P is a necessary and sufficient condition for Q".
"P if and only if Q".
"P iff Q".

The truth tables for logical operators are used to determine the truth-values of arbitrary propositional forms. Whenever there are n propositional variables in a propositional form, we have 2^n possible combinations of true and false to investigate.

Example 2.6 The truth table for the propositional form $\neg(P \wedge \neg Q)$ is constructed as

P	Q	$\neg Q$	$P \wedge \neg Q$	$\neg(P \wedge \neg Q)$
0	0	1	0	1
0	1	0	0	1
1	0	1	1	0
1	1	0	0	1

We see that for two variables we have to investigate four different cases.

2.3   Types of Propositional Forms

In propositional logic, we have special cases where a propositional form is either always true or always false, regardless of the truth-values of the propositional variables.

Definition 2.4 (Tautology, Contradiction, and Contingency). A propositional form whose truth-value is true for all possible truth-values of its propositional variables is called a *tautology*. A *contradiction* (or *absurdity*) is a propositional form that is always false. A *contingency* is a propositional form that is neither a tautology nor a contradiction.

The following examples illustrate the concepts of tautology, contradiction, and contingency.

Example 2.7 The propositional form $(P \wedge Q) \Rightarrow P$ is a tautology.

P	Q	$P \wedge Q$	$(P \wedge Q) \Rightarrow P$
0	0	0	1
0	1	0	1
1	0	0	1
1	1	1	1

Example 2.8 The propositional form $P \wedge \neg P$ is a contradiction.

P	$\neg P$	$P \wedge \neg P$
0	1	0
1	0	0

This propositional form corresponds to the law of contradiction which states that something cannot be true and false at the same time. Another fundamental law of logic is the law of the excluded middle (also called "tertium non datur" in Latin for "a third one is not given") or $P \vee \neg P$, stating that something can only be true or false and there is nothing in between.

Example 2.9 The propositional form $(P \vee \neg Q) \Rightarrow Q$ is a contingency.

P	Q	$\neg Q$	$P \vee \neg Q$	$(P \vee \neg Q) \Rightarrow Q$
0	0	1	1	0
0	1	0	0	1
1	0	1	1	0
1	1	0	1	1

Definition 2.5 (Logical Identity). Two propositional forms $\Phi(P, Q, R, \ldots)$ and $\Psi(P, Q, R, \ldots)$ are said to be *logically equivalent* when their truth tables are identical, or when the equivalence $\Phi(P, Q, R, \ldots) \Leftrightarrow \Psi(P, Q, R, \ldots)$ is a tautology. Such equivalence is also called a *logical identity*.

We can replace one propositional form with its equivalent form. This often helps to simplify logical expressions. Table 2.3 lists the most important logical identities.

In Table 2.3, **1** and **0** denote propositions that are always true or always false, respectively. Identity (18) allows us to replace the implication by negation and disjunction. The equivalence can be replaced by implications through identity (19). Identities (7) and (8) (De Morgan's Laws) allow the replacement of conjunction by disjunction and vice versa. All the identities can be proven by constructing their truth tables using the truth tables of the logical operators presented in Table 2.2.

Example 2.10 Simplify the following propositional form: $\neg(\neg P \Rightarrow \neg Q)$.

The numbers on the right indicate the identities that have been applied to simplify the propositional form:

$$
\begin{array}{ll}
\neg(\neg P \Rightarrow \neg Q) & (22) \\
\neg(Q \Rightarrow P) & (18) \\
\neg(\neg Q \vee P) & (7) \\
\neg\neg Q \wedge \neg P & (17) \\
Q \wedge \neg P & (4) \\
\neg P \wedge Q &
\end{array}
$$

Many useful tautologies are implications. Table 2.4 lists the most important ones of them.

Table 2.3 Logical identities

(1)	$P \Leftrightarrow (P \vee P)$	Idempotence of $\vee$
(2)	$P \Leftrightarrow (P \wedge P)$	Idempotence of $\wedge$
(3)	$(P \vee Q) \Leftrightarrow (Q \vee P)$	Commutativity of $\vee$
(4)	$(P \wedge Q) \Leftrightarrow (Q \wedge P)$	Commutativity of $\wedge$
(5)	$[(P \vee Q) \vee R] \Leftrightarrow [P \vee (Q \vee R)]$	Associativity of $\vee$
(6)	$[(P \wedge Q) \wedge R] \Leftrightarrow [P \wedge (Q \wedge R)]$	Associativity of $\wedge$
(7)	$\neg(P \vee Q) \Leftrightarrow (\neg P \wedge \neg Q)$	De Morgan's Laws
(8)	$\neg(P \wedge Q) \Leftrightarrow (\neg P \vee \neg Q)$	
(9)	$[P \wedge (Q \vee R)] \Leftrightarrow [(P \wedge Q) \vee (P \wedge R)]$	Distributivity of $\wedge$ over $\vee$
(10)	$[P \vee (Q \wedge R)] \Leftrightarrow [(P \vee Q) \wedge (P \vee R)]$	Distributivity of $\vee$ over $\wedge$
(11)	$(P \vee \mathbf{1}) \Leftrightarrow \mathbf{1}$	
(12)	$(P \wedge \mathbf{1}) \Leftrightarrow P$	
(13)	$(P \vee \mathbf{0}) \Leftrightarrow P$	
(14)	$(P \wedge \mathbf{0}) \Leftrightarrow \mathbf{0}$	
(15)	$(P \vee \neg P) \Leftrightarrow \mathbf{1}$	Law of the excluded middle
(16)	$(P \wedge \neg P) \Leftrightarrow \mathbf{0}$	Law of contradiction
(17)	$P \Leftrightarrow \neg(\neg P)$	Double negation
(18)	$(P \Rightarrow Q) \Leftrightarrow (\neg P \vee Q)$	Implication
(19)	$(P \Leftrightarrow Q) \Leftrightarrow [(P \Rightarrow Q) \wedge (Q \Rightarrow P)]$	Equivalence
(20)	$[(P \wedge Q) \Rightarrow R] \Leftrightarrow [P \Rightarrow (Q \Rightarrow R)]$	Exportation
(21)	$[(P \Rightarrow Q) \wedge (P \Rightarrow \neg Q)] \Leftrightarrow \neg P$	Absurdity
(22)	$(P \Rightarrow Q) \Leftrightarrow (\neg Q \Rightarrow \neg P)$	Contrapositive

Table 2.4 Logical implications

(1)	$P \Rightarrow (P \vee Q)$	Addition
(2)	$(P \wedge Q) \Rightarrow P$	Simplification
(3)	$[P \wedge (P \Rightarrow Q)] \Rightarrow Q$	Modus ponens
(4)	$[(P \Rightarrow Q) \wedge \neg Q] \Rightarrow \neg P$	Modus tollens
(5)	$[\neg P \wedge (P \vee Q)] \Rightarrow Q$	Disjunctive syllogism
(6)	$[(P \Rightarrow Q) \wedge (Q \Rightarrow R)] \Rightarrow (P \Rightarrow R)$	Hypothetical syllogism
(7)	$(P \Rightarrow Q) \Rightarrow [(Q \Rightarrow R) \Rightarrow (P \Rightarrow R)]$	
(8)	$[(P \Rightarrow Q) \wedge (R \Rightarrow S)] \Rightarrow [(P \wedge R) \Rightarrow (Q \wedge S)]$	
(9)	$[(P \Leftrightarrow Q) \wedge (Q \Leftrightarrow R)] \Rightarrow (P \Leftrightarrow R)$	

Some of these implications correspond to rules of inference (or logical rules) that will be discussed later.

2.4   Applications in GIS

In GIS applications, we find logical operators mainly in spatial analysis and database queries. Figure 2.1 shows the Raster Calculator of ArcGIS Pro Spatial Analyst and its logical connectors ("and", "or", "xor", and "not"). In this example, all raster cells with an elevation between 1,000 and 1,500 will be selected. The logical connector "and" is represented by the character "&".

The logical implication can be found in every programming language in the form of the *if-statement*, which takes the general form (in pseudo code) as follows:

```
if <condition> then <statement> else <statement>
```

The condition contains an expression that can be evaluated as either true or false (proposition). Logical connectors or comparison operators are often part of the condition. The following Python code sample checks whether a data set in a geodatabase exists and deletes it if it exists.

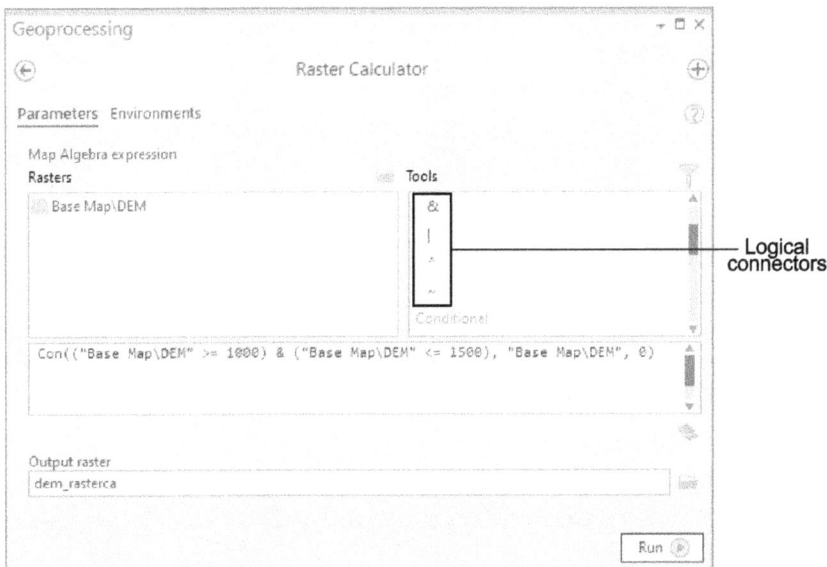

Figure 2.1   Raster Calculator with logical connectors

```
import arcpy
arcpy.env.workspace = "C:/Data/database.gdb"
if arcpy.Exists("waterbodies"):
  arcpy.Delete_managemnt("waterbodies")
else:
  print("waterbodies does not exist.")
```

2.5   Exercises

Exercise 2.1   Construct the truth table of the propositional form $[(P \Leftrightarrow Q) \wedge Q] \Rightarrow P$.

Exercise 2.2   Show that $(P \wedge Q) \Rightarrow (P \vee Q)$ is a tautology.

Exercise 2.3   Simplify the propositional form $\neg(P \vee Q) \vee (\neg P \wedge Q)$.

Exercise 2.4   Let P be the proposition "It is raining". Let Q be the proposition "I get wet". Let R be the proposition "I am sick".
(1) Write the following propositions in symbolic notation:
    (i)   "If it is raining, then I get wet and I am sick".
    (ii)   "I am sick if it is raining".
    (iii)   "I do not get wet".
    (iv)   "It is raining and I am not sick".
(2) Express the following propositional forms in plain English:
    (i) $R \wedge Q$.
    (ii) $(P \Rightarrow Q) \vee \neg R$.
    (iii) $\neg(R \vee Q)$.
    (iv) $(Q \Rightarrow R) \wedge (R \Rightarrow Q)$.

Exercise 2.5   Write down the converse and contrapositive of the following propositional forms:
(1) "If it rains, then I get wet".
(2) "I will stay only if he leaves".
(3) "I will not pass the exam, if I do not study hard".

Exercise 2.6 For the following expressions, find equivalent expressions using identities. The equivalent expressions must use only the operators $\land$ and $\neg$, and must be as simple as possible.

(1) $P \lor Q \lor \neg R$.

(2) $P \lor [(\neg Q \land R) \Rightarrow P]$.

(3) $P \Rightarrow (Q \Rightarrow P)$.

Exercise 2.7 In a computer program you have the following statement $x \leftarrow y$ and FUNC(y, z), where x, y are logical variables, FUNC is a logical function and z is an output variable. The value of z is determined by the execution of the function FUNC. Optimizing compilers generate code that is only executed when really needed. Assume such an optimized code has been generated for your program. Can you always rely on that a value for z is computed?

Chapter 3

Predicate Logic

The language of propositional logic is not powerful enough to make all the assertions needed in mathematics. We frequently need to make general statements about the properties of an object or relationships between objects, such as "All humans are mortal" or the equation "$x + y = 2$".

This chapter introduces the concepts of predicates and quantifiers that enrich the language of logic and allow making assertions in a much more general way than what is possible in propositional logic. The knowledge acquired about predicates will be used to translate natural language statements into the form of predicates.

3.1 Predicates

In propositional logic, we cannot make assertions such as "$x + y = 5$" or "$x \leqslant y$", because the truth-values of these statements depend on the values of the variables x and y. Only when we assign values to the variables, the assertions become propositions.

We also make assertions in natural languages like "Ann lives in Vienna" or "All humans are mortal" that correspond to a general construct "x lives in y" or "for all x, $M(x)$". These constructs express a relationship between objects or a property of objects.

Definition 3.1 (Predicate). A term designating a property or relationship is called a *predicate*.

Assertions made with predicates and variables become propositions when the variables are replaced by specific values.

Example 3.1   In the assertion "x lives in y" x and y are variables, and "lives in" is a predicate. When we replace x by "Ann" and y by "Vienna" it becomes the proposition "Ann lives in Vienna".

Example 3.2   Predicates appear commonly in computer programs as control statements of high-level programming languages. The statement "if $x < 5$ then $y \leftarrow 2 * y$" for instance contains the predicate "$x < 5$". When the program runs the current value of x determines the truth-value of "$x < 5$".

Some predicates have a well-known notation in mathematics. Examples are "equal to" and "greater than", usually written as "=" and ">", respectively. Otherwise, we will denote predicates with upper case letters.

Example 3.3   The assertion "x is a woman" could be written as $W(x)$, "x lives in y" could be written as $L(x, y)$, and "$x + y = z$" could be written as $S(x, y, z)$.

Definition 3.2 (Variables, Universe).   In the expression $P(x_1, x_2, \ldots, x_n)$, P is a predicate,[1] and x_i are variables. When P has n variables we say that it has *n arguments* or it is an *n-place predicate*. Values for the variables must come from a set called the *universe of discourse*, or the *universe*. The universe is normally denoted as U and must contain at least one element.

When we take values $c_1, c_2, \ldots, c_n$ from the universe and assign them to the variables of a predicate $P(x_1, x_2, \ldots, x_n)$, we get a proposition $P(c_1, c_2, \ldots, c_n)$.

Definition 3.3 (Valid, Satisfiable).   If $P(c_1, c_2, \ldots, c_n)$ is true for every choice of elements from the universe, then we say that P is *valid in the universe U*. If $P(c_1, c_2, \ldots, c_n)$ is true for some elements of the universe, we say that P is *satisfiable in the universe U*. The values $c_1, c_2, \ldots, c_n$ that make $P(c_1, c_2, \ldots, c_n)$ true are said to *satisfy P*. If $P(c_1, c_2, \ldots, c_n)$ is false for every choice of values of the universe we say that P is *unsatisfiable in U*.

---

[1]To be precise, we must distinguish between *predicate variables* and *predicate constants*. Whenever we use specific predicates, such as W, L or S in Example 3.3, we actually deal with predicate constants, whereas an expression like P with no immediate interpretation of the predicate denotes a predicate variable.

3.2 Quantifiers

We have seen that a predicate can become a proposition by substituting values for the arguments. We say that the variables are *bound*. There are two ways of binding variables of predicates.

Definition 3.4 (Binding of Variables). Variables of predicates can be bound by assigning values to them, or by quantifying them. We know two *quantifiers*, the *universal quantifier* and the *existential quantifier*.

If $P(x)$ is a predicate then the assertion "for all x, $P(x)$" (which means, "for all values of x, the assertion $P(x)$ is true") is a statement in which the variable x is *universally quantified*. The universal quantifier "for all" is written as $\forall$, and can be read as "for all", "for every", "for any", "for arbitrary", or "for each". The statement "for all x, $P(x)$" becomes "$\forall x P(x)$". We say that $\forall x P(x)$ is true if and only if $P(x)$ is valid in U; otherwise, it is false.

If $P(x)$ is a predicate then the assertion "for some x, $P(x)$" (which means, "there exists at least one value of x for which the assertion $P(x)$ is true") is a statement in which the variable x is *existentially quantified*. The existential quantifier "there exists" is written as $\exists$, and can be read as "there exists", "for some" or "for at least one". The statement "for some x, $P(x)$" becomes "$\exists x P(x)$". We say that $\exists x P(x)$ is true if and only if $P(x)$ is satisfiable in U; otherwise, it is false.

There is a variation of the existential quantifier to assert that there is one and only one element in the universe, which makes a predicate true. This quantifier is read as "there is one and only one x such that ...", "there is exactly one x such that ..." or "there is a unique x such that ...". It is written as $\exists!$.

Example 3.4 Let us assume the universe to be all integers $\mathbb{Z}$ and the following propositions are formed by quantification:
(1) $\forall x[x - 1 < x]$.
(2) $\forall x[x = 5]$.
(3) $\forall x \forall y[x + y > x]$.
(4) $\exists x[x < x + 1]$.
(5) $\exists x[x = 5]$.
(6) $\exists x[x = x + 1]$.
(7) $\exists! x[x = 5]$.

Propositions (1), (4), (5), and (7) are true. Proposition (3) is false in the integers; however, it would be true in the positive integers $\mathbb{Z}^+$. Propositions (2) and (6) are false.

As we have seen above variables can be bound by assigning values to them. We can also express quantified assertions with propositions by assigning all elements of the universe to the variables and combining them with logical operators.

Definition 3.5 (Propositional Form of Quantifiers). If the universe U consists of the elements $c_1, c_2, c_3, \ldots$, then the propositions $\forall x P(x)$ and $\exists x P(x)$ can be written as $P(c_1) \wedge P(c_2) \wedge P(c_3) \wedge \ldots$ and $P(c_1) \vee P(c_2) \vee P(c_3) \vee \ldots$, respectively.

All variables must be bound to transform a predicate into a proposition. If in an n-place predicate, m variables are bound, we say that the predicate has $n - m$ *free variables*.

Example 3.5 The predicate $P(x, y, z)$ representing "$x + y < z$" has three variables. If we bind one variable, e.g., x is assigned the value 2, then we get the predicate $P(2, y, z)$ with two free variables, representing "$2 + y < z$".

The order in which the variables are bound is the same as the order in the quantifier list when more than one quantifier is applied to a predicate. Therefore $\forall x \forall y P(x, y)$ has to be evaluated as $\forall x [\forall y P(x, y)]$. The order of the quantifiers is not arbitrary. It affects the meaning of an assertion. $\forall x \exists y$ has not the same meaning as $\exists y \forall x$. The only exception is that we can always replace $\forall x \forall y$ by $\forall y \forall x$, and $\exists x \exists y$ by $\exists y \exists x$.

Example 3.6 If $P(x, y)$ denotes the predicate "x is the child of y" in the universe of all persons. Then the proposition $\forall x \exists y P(x, y)$ means, "Everyone is the child of someone", whereas $\exists y \forall x P(x, y)$ means, "There is a person so that everyone is the child of this person".

3.3 Quantifiers and Logical Operators

When we express mathematical or natural language statements, we generally need quantifiers, predicates and logical operators. These statements can take on a variety of different forms.

Example 3.7 Let the universe be the integers and $E(x)$ denote "x is an even number", $O(x)$ denote "x is an odd number", $N(x)$ denote "x is a non-negative integer", and $P(x)$ denote "x is a prime number". The following examples show how assertions can be expressed in the language of predicate logic.

(1) There exists an odd integer. $\exists x O(x)$
(2) Every integer is even or odd. $\forall x[E(x) \lor O(x)]$
(3) All prime numbers are non-negative. $\forall x[P(x) \Rightarrow N(x)]$
(4) The only even prime number is two. $\forall x[(E(x) \land P(x)) \Rightarrow x = 2]$
(5) There is only one even prime number. $\exists! x[E(x) \land P(x)]$
(6) Not all prime numbers are odd. $\neg \forall x[P(x) \Rightarrow O(x)]$, or $\exists x[P(x) \land \neg O(x)]$
(7) If an integer is not even, then it is odd. $\forall x[\neg E(x) \Rightarrow O(x)]$

In analogy to tautologies, contradictions and contingencies in propositional logic we can also establish types of assertions involving predicate variables.[2]

Definition 3.6 (Validity of Assertions with Predicate Variables). An assertion involving predicate variables is *valid* if it is true for every universe. An assertion is *satisfiable* if there exist a universe and some interpretations of the predicate variables that make it true. It is *unsatisfiable* if there is no universe and no interpretation that make the assertion true. Two assertions A_1 and A_2 are *logically equivalent* if for every universe and every interpretation of the predicate variables $A_1 \Leftrightarrow A_2$, i.e., A_1 is true iff A_2 is true.

The *scope* of a quantifier is the part of the assertion for which variables are bound by this quantifier.

Example 3.8 In the assertion $\forall x[P(x) \land Q(x)]$ the scope of the universal quantifier is $P(x) \land Q(x)$. In the assertion $[\exists x P(x)] \Rightarrow [\forall x Q(x)]$ the scope of $\exists$ is $P(x)$ and the scope of $\forall$ is $Q(x)$.

Table 3.1 shows a list of logical equivalences and other relationships between assertions involving quantifiers.

The logical equivalencies (3) and (5) can be used to propagate negation signs through a sequence of quantifiers.

---

[2]For the notion of predicate variable, see the footnote for Definition 3.2.

Table 3.1   Logical relationships involving quantifiers

(1)	$\forall x P(x) \Rightarrow P(c)$, where c is an arbitrary element of the universe
(2)	$P(c) \Rightarrow \exists x P(x)$, where c is an arbitrary element of the universe
(3)	$\forall x \neg P(x) \Leftrightarrow \neg \exists x P(x)$
(4)	$\forall x P(x) \Rightarrow \exists x P(x)$
(5)	$\exists x \neg P(x) \Leftrightarrow \neg \forall x P(x)$
(6)	$[\forall x P(x) \wedge Q] \Leftrightarrow \forall x [P(x) \wedge Q]$
(7)	$[\forall x P(x) \vee Q] \Leftrightarrow \forall x [P(x) \vee Q]$
(8)	$[\exists x P(x) \wedge Q] \Leftrightarrow \exists x [P(x) \wedge Q]$
(9)	$[\exists x P(x) \vee Q] \Leftrightarrow \exists x [P(x) \vee Q]$
(10)	$[\forall x P(x) \wedge \forall x Q(x)] \Leftrightarrow \forall x [P(x) \wedge Q(x)]$
(11)	$[\forall x P(x) \vee \forall x Q(x)] \Rightarrow \forall x [P(x) \vee Q(x)]$
(12)	$\exists x [P(x) \wedge Q(x)] \Rightarrow [\exists x P(x) \wedge \exists x Q(x)]$
(13)	$[\exists x P(x) \vee \exists x Q(x)] \Leftrightarrow \exists x [P(x) \vee Q(x)]$

Equivalencies (6), (7), (8), and (9) tell us that whenever a proposition occurs within the scope of a quantifier, it can be removed from the scope of the quantifier. Predicates whose variables are not bound by a quantifier can also be removed from the scope of this quantifier.

Statements (10) and (12) show that the universal quantifier *distributes* over the conjunction, but the existential quantifier does not. (11) and (13) show that the existential quantifier distributes over the disjunction, but the universal quantifier does not.

3.4   Compact Notation

The form of logical notation as presented here is often too complex to express relatively simple assertions in mathematical language. Therefore, a compact form of logical notation is used.

For the assertion "for every x such that $x \geqslant 0$, $P(x)$ is true" we would have to write $\forall x [(x \geqslant 0) \Rightarrow P(x)]$. Instead we can write in compact notation $\forall x_{x \geq 0} P(x)$. In the same way, we write for the assertion "there exists an x such that $x \neq 5$ and $P(x)$ is true" $\exists x [(x \neq 5) \wedge P(x)]$ in the long notation and $\exists x_{x \neq 5} P(x)$ in the compact notation. This notation also allows to propagate the negation sign through quantifiers as mentioned in logical equivalencies (3) and (5) of Table 3.1.

3.5 Applications in GIS

In relational database technology, we use the select operator to select a subset of tuples t (or records) in a relation that satisfies a given selection condition. In general, we can denote the select operator as $\sigma_{\text{selection condition}}$ (relation name) or $\sigma_{\phi(t)}(R)$ when we substitute selection condition with $\phi(t)$ and R for the relation name. The selection condition is a predicate, i.e., it designates a property of the tuples, and we can thus write the general selection as a predicative set expression $\{t \in R | \phi(t)\}$.

Let ARC(ID,StartNode,EndNode,LPoly,RPoly) be a relation schema describing arcs in a topologically structured data set. The selection operator $\sigma_{(\text{LPoly} = \text{"A" OR RPoly} = \text{"A"})}(\text{ARC})$ results in all arcs that form the boundary of polygon A. A translation of this selection into standard SQL reads as

```
SELECT * FROM ARC WHERE LPoly = 'A' or RPoly = 'A'.
```

3.6 Exercises

Exercise 3.1 Translate the following assertions into the notation of predicate logic (the universe is given in parentheses):
(1) "If 3 is odd, some numbers are odd". (integers)
(2) "Some cats are blue". (animals)
(3) "All cats are blue". (animals)
(4) "There are areas, lines, and points". (geometric figures)
(5) "If x is greater than y and y is greater than z, then x is greater than z". (integers)
(6) "When it is night all cats are black". (animals)
(7) "When it is daylight some cats are black". (animals)
(8) "All students of this course are happy if they pass the mathematics exam". (all students)

Chapter 4

Logical Inference

This chapter introduces the concept of logical arguments and rules of inference (or logical rules). Starting from a set of premises (or hypotheses) a conclusion is drawn. If the conclusion follows logically from the premises, the argument is valid. If this is not the case then the conclusion cannot be drawn from the hypotheses. One basic rule of inference, the rule of deduction, is the foundation of rule-based systems.

In a formal mathematical system, we assume a set of axioms that are a set of given true statements. From these axioms, we derive assertions that can be shown to be true. These assertions are called theorems. A proof is an argument, which establishes the truth of a theorem.

4.1 Logical Arguments

Often, we assume that certain assumptions are true, and we draw a conclusion from these assumptions. If, for instance, we assume that the two statements "It is raining" and "If it is raining, then I get wet" are true, then we can conclude that "I get wet".

Definition 4.1 (Logical Argument). A *logical argument* consists of a set of *hypotheses* (or *premises*) that are assumed to be true. The *conclusion* follows from the premises. *Rules of inference* specify which conclusions can be drawn from assertions known or assumed to be true. An argument is said to be *valid* (or *correct*) when the conclusion follows logically from the premises.

Logical arguments are usually written in the form of

P_1

P_2

$\vdots$

$\dfrac{P_n}{\therefore Q}$

where P_i are the premises and Q is the conclusion.

Example 4.1   The argument presented above is written as

P1:   It is raining.

P2:   If it is raining, then I get wet.

Conclusion: I get wet.

The rule of inference applied is of the form

P

$\dfrac{P \Rightarrow Q}{\therefore Q}$

4.2   Proving Arguments Valid in Propositional Logic

In general, arguments can be proven valid in two ways, using truth tables or using rules of inference. In the first case, an argument has to be translated into its equivalent tautological form. The procedure is straightforward:

(1) Identify all propositions.

(2) Assign propositional variables to the propositions.

(3) Write the argument in its tautological form using the propositional variables.

(4) Evaluate the tautological form using a truth table.

Note that the more propositions are involved the more tedious the procedure becomes. In the case of applying rules of inference, the trick is to find the right rules and apply them properly.

4.2.1   Proving Arguments Valid with Truth Tables

Every logical argument with n premises $P_1, P_2, \ldots, P_n$ and the conclusion Q can be written as a propositional form $(P_1 \wedge P_2 \wedge \cdots \wedge P_n) \Rightarrow Q$. If this propositional form is a tautology, the argument is correct.

Example 4.2 The argument in Example 4.1 contains the propositions P "It is raining" and Q "I get wet". It has the tautological form $[P \wedge (P \Rightarrow Q)] \Rightarrow Q$. The proof that this is a tautology is left to the reader.

4.2.2 Proving Arguments Valid with Rules of Inference

The second way to prove an argument valid is to apply rules of inference. They are applied to the premises until the conclusion follows (argument valid) or the conclusion cannot be reached (argument invalid). Table 4.1 shows the most important rules of inference, their tautological forms and the names that are given to them by logicians.

Table 4.1 Rules of inference

Rule of inference	Tautological form	Name
$\dfrac{P}{\therefore P \vee Q}$	$P \Rightarrow (P \vee Q)$	Addition
$\dfrac{P \wedge Q}{\therefore P}$	$(P \wedge Q) \Rightarrow P$	Simplification
$\dfrac{\begin{array}{c}P \\ P \Rightarrow Q\end{array}}{\therefore Q}$	$[P \wedge (P \Rightarrow Q)] \Rightarrow Q$	Modus ponens
$\dfrac{\begin{array}{c}\neg Q \\ P \Rightarrow Q\end{array}}{\therefore \neg P}$	$[\neg Q \wedge (P \Rightarrow Q)] \Rightarrow \neg P$	Modus tollens
$\dfrac{\begin{array}{c}P \vee Q \\ \neg P\end{array}}{\therefore Q}$	$[(P \vee Q) \wedge \neg P] \Rightarrow Q$	Disjunctive Syllogism
$\dfrac{\begin{array}{c}P \Rightarrow Q \\ Q \Rightarrow R\end{array}}{\therefore P \Rightarrow R}$	$[(P \Rightarrow Q) \wedge (Q \Rightarrow R)] \Rightarrow [P \Rightarrow R]$	Hypothetical Syllogism
$\dfrac{\begin{array}{c}P \\ Q\end{array}}{\therefore P \wedge Q}$	$(P \wedge Q) \Rightarrow (P \wedge Q)$	Conjunction
$\dfrac{\begin{array}{c}(P \Rightarrow Q) \wedge (R \Rightarrow S) \\ P \vee R\end{array}}{\therefore Q \vee S}$	$[(P \Rightarrow Q) \wedge (R \Rightarrow S) \wedge (P \vee R)] \Rightarrow [Q \vee S]$	Constructive Dilemma
$\dfrac{\begin{array}{c}(P \Rightarrow Q) \wedge (R \Rightarrow S) \\ \neg Q \vee \neg S\end{array}}{\therefore \neg P \vee \neg R}$	$[(P \Rightarrow Q) \wedge (R \Rightarrow S) \wedge (\neg Q \vee \neg S)] \Rightarrow [\neg P \vee \neg R]$	Destructive Dilemma

Some of these rules of inference are evident. The disjunctive syllogism, for instance, simply says that if you have two options and one is not available, then you choose the other one.[1]

Example 4.3   The argument presented in Example 4.1 is a straightforward application of the *modus ponens*.

Example 4.4   In the same way as above the argument "I do not get wet", and "If it is raining, then I get wet", therefore "It is not raining" is a straightforward application of the *modus tollens*.

4.3   Proving Arguments Valid in Predicate Logic

When we want to prove the validity of an argument that contains predicates and quantifiers, we need more rules. Table 4.2 shows some of the rules of inference involving predicates and quantifiers.

Table 4.2   Rules of inference involving predicates and quantifiers

Rule of inference	Name
$\dfrac{\forall x P(x)}{\therefore P(c)}$	Universal instantiation
$\dfrac{P(x)}{\therefore \forall x P(x)}$	Universal generalization
$\dfrac{\exists x P(x)}{\therefore P(c)}$	Existential instantiation
$\dfrac{P(c)}{\therefore \exists x P(x)}$	Existential generalization

The universal instantiation allows us to conclude from the fact that if a predicate is valid in a given universe, then it is also valid for one individual from that universe. The universal generalization permits us to conclude that if we can prove that a predicate is valid for every element of the given universe, then the universally quantified assertion holds.

---

[1]Most people would agree with that and even dogs or cats know that.

The existential instantiation concludes from the truth that if there is at least one element of the universe for which the predicate is true, then there is one element c for which $P(c)$ is true. The existential generalization allows us to conclude from the truth that if a predicate is true for one particular element of the universe, then the existentially quantified assertion $\exists x P(x)$ is true.

Example 4.5 Let us consider the following argument:

P1:   Every man has a brain.

P2:   John Williams is a man.

Conclusion: Therefore, John Williams has a brain.

Let $M(x)$ denote the assertion "x is a man", $B(x)$ denote the assertion "x has a brain", and W denote John Williams. Then the logical argument can be expressed as:

(1) $\forall x[M(x) \Rightarrow B(x)]$.

(2) $M(W)$.

(3) $\therefore B(W)$.

A formal proof of the argument is shown in Table 4.3.

Table 4.3   Formal proof of the argument

Assertion	Reason
(1) $\forall x[M(x) \Rightarrow B(x)]$	Hypothesis 1
(2) $M(W) \Rightarrow B(W)$	Step (1) and universal instantiation
(3) $M(W)$	Hypothesis 2
(4) $B(W)$	Steps (2) and (3) and modus ponens

We do not go deeper into the theory of proving arguments in general predicate logic.

4.4   Applications in GIS

Rule-based systems apply rules to data provided using an inference engine. These systems are also called expert systems, and are widely applied in the geosciences. Spatial decision support systems (SDSS) are rule-based systems that are designed and tuned for spatial data.

Rules are stored as implications in the form "if <premise> then <consequence>". The inference engine examines the given data in the database and determines if they match a given premise. If this is the case, the consequence is applied accordingly. This is a straightforward application of the *modus ponens*.

4.5 Exercises

Exercise 4.1 Translate the following argument into a symbolic notation and check if it is correct:
P1:   If I study then I do not fail the mathematics exam.
P2:   If I do not play soccer then I study.
P3:   I fail the mathematics exam.
Conclusion: I play soccer.

Exercise 4.2 Translate the following argument into a symbolic notation and check if it is correct using a truth table:
P1:   If the Earth is a disk then I do not reach the USA.
P2:   If I travel west then I reach the USA.
P3:   I do not travel west and I reach the USA.
Conclusion: The Earth is not a disk.

Exercise 4.3 Translate the following argument into a symbolic notation and check if it is correct:
P1:   If 6 is not an even number, then 5 is not a prime number.
P2:   6 is an even number.
Conclusion: 5 is a prime number.

Chapter 5

Set Theory

Sets are the very fundamental building block of many mathematical theories. They are intuitively perceived as a collection of well-distinguished objects. The formal definition and axiomatic foundation of set theory are more complicated and will not be discussed here.

Starting from an intuitive definition of sets, we explore relations between sets and operations on sets. The fundamental principles of subset and set equality as well as set union, intersection, and difference are explained.

5.1 Sets and Elements

Set theory was developed by Georg Cantor (1845–1918) as what we call today naive set theory. This is a more intuitive approach than the axiomatic set theory. However, in naive set theory there is the possibility for logical contradictions (or paradoxes), something that should not occur in a formal system.

Definition 5.1 (Set). A set is a collection of well-distinguished objects. Any object of the collection is called an element or a member of the set. An element x of a set S is written as $x \in S$. If x is not a member of S, we write $x \notin S$. If a set has a finite number of elements we call it a finite set. A set with no elements is called the empty set, null set or void set and is denoted as $\{\ \}$ or $\emptyset$.

There are many ways to specify a set. A finite set can be specified explicitly by listing all its elements. The set A consisting of the natural numbers smaller than 10 can be written as $A = \{1, 2, 3, 4, 5, 6, 7, 8, 9\}$,[1] or we can describe the

---

[1] In this book the natural numbers do not contain 0.

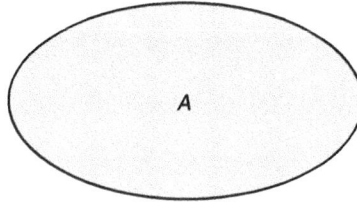

Figure 5.1   Venn diagram

set implicitly by means of a predicate and a free variable, $A = \{x | x \in \mathbb{N} \wedge x < 10\}$. We can also draw a set with a Venn diagram (Figure 5.1).

In order to indicate the "size" of a set we need a measure. This is defined as the number of elements (or cardinality).

Definition 5.2 (Cardinality). The *cardinality* of a set S is the number of its elements, written as $|S|$.

Example 5.1   The set $A = \{x | x$ is a character of the English alphabet$\}$ has the cardinality $|A| = 26$.

Example 5.2   The natural numbers $\mathbb{N}$ are an infinite set. Their cardinality is denoted as $\aleph_0$ (pronounced as aleph zero[2]). Every set S that has the same cardinality as the natural numbers is called a *countably infinite set*.[3] The proof is usually established by finding a one-to-one function that maps the natural numbers to S. The cardinality of the integers $\mathbb{Z}$ and the rational numbers $\mathbb{Q}$ is $\aleph_0$. The rational numbers $\mathbb{Q}$ are all fractions of the form $\frac{a}{b}$ where $a, b \in \mathbb{Z}$ and $b \neq 0$. The cardinality of the real numbers $\mathbb{R}$ is denoted as c (the continuum). They are said to be *uncountably infinite*. There are more real numbers than rational numbers, and there are more rational numbers than integers.

Example 5.3   The cardinality of the set $A = \{1, 1, 2, 2, 2, 3\}$ is 3, because the elements of a set must be distinguishable. In A, element 1 appears twice, 2 appears three times, and 3 appears once. Since it does not matter how often an element is repeated, the number of elements is three.

---

[2] Aleph is the first character in the Hebrew alphabet.

[3] A set is *finite* if there exists a one to one correspondence between its elements and a subset of the natural numbers $\{1, 2, 3, \ldots, n\}$ for some n (including $n = 0$ for the empty set). A set is *countable* if it is either finite or countably infinite.

Example 5.4 The set $A = \{\emptyset\}$ has one element, the empty set. Therefore, its cardinality is 1. Although the empty set has cardinality zero, here the empty set appears as an element of set A.

5.2 Relations between Sets

We know two relations between sets, subset and equality. The subset relation refers to the containment of one set in another.

Definition 5.3 (Subset). If each element of a set A is an element of a set B then A is *subset* of B, written as $A \subseteq B$. B is called *superset* of A, written as $B \supseteq A$. We call a set A a proper subset of B when $A \subset B$ and $A \neq B$.

Two sets A and B are equal and written as $A = B$ if and only if $A \subset B$ and $B \subset A$.

   The following statements can be derived from the definitions of sets and their relationships:

(1) If U is the universe of discourse, then $A \subseteq U$.

(2) For any set A, $A \subseteq A$.

(3) If $A \subseteq B$ and $B \subseteq C$, then $A \subseteq C$. The empty set is subset of every set, or for any set A, $\emptyset \subseteq A$.

5.3 Operations on Sets

In the following, we consider operations on sets that use given sets (*operands*) to produce a new set (*resultant*).

Definition 5.4 (Union). The *union* of two sets A and B, written as $A \cup B$ is the set $A \cup B = \{x | x \in A \vee x \in B\}$.

Definition 5.5 (Intersection). The *intersection* of two sets A and B, written as $A \cap B$ is the set $A \cap B = \{x | x \in A \wedge x \in B\}$. If $A \cap B = \emptyset$, we say that the two sets are *disjoint*.

Definition 5.6 (Difference). The *difference* of two sets A and B, written as $A - B$ (or $A \backslash B$) is the set $A - B = \{x | x \in A \wedge x \notin B\}$.

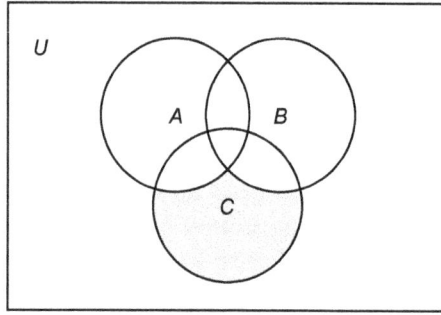

Figure 5.2   Venn diagram of the operations $\overline{A \cup B} \cap C$.

Definition 5.7 (Complement). The *complement* of a set A, written as $\overline{A}$, is the set $\overline{A} = U - A = \{x | x \notin A\}$, where U is the universe of discourse.

Example 5.5   The Venn diagram in Figure 5.2 illustrates the operations $\overline{A \cup B} \cap C$.

Union and intersection can generally be defined for more than two sets. Let I be an arbitrary finite or infinite index set. Every element $i \in I$ has assigned a set A_i, then the *union* of the A_i is defined as

$$\bigcup_{i \in I} A_i = \{x | \exists i [i \in I \wedge x \in A_i]\}$$

In the same way, we define the *intersection* of the A_i as

$$\bigcap_{i \in I} A_i = \{x | \forall i [i \in I \Rightarrow x \in A_i]\}$$

If we know the number of elements n in the index set, we write

$$\bigcup_{i=1}^{n} A_i$$

and

$$\bigcap_{i=1}^{n} A_i$$

respectively.

Table 5.1 summarizes some of the most important rules for set operations. They can be easily proven by translating them into their equivalent forms in the language of logic and show that they are tautologies.

Table 5.1 Rules for set operations

(1)	$A \cup A = A$	
(2)	$A \cap A = A$	
(3)	$(A \cup B) \cup C = A \cup (B \cup C)$	associativity
(4)	$(A \cap B) \cap C = A \cap (B \cap C)$	
(5)	$A \cup B = B \cup A$	commutativity
(6)	$A \cap B = B \cap A$	
(7)	$A \cup (B \cap C) = (A \cup B) \cap (A \cup C)$	distributivity
(8)	$A \cap (B \cup C) = (A \cap B) \cup (A \cap C)$	
(9)	$\overline{A \cup B} = \overline{A} \cap \overline{B}$	De Morgan's laws
(10)	$\overline{A \cap B} = \overline{A} \cup \overline{B}$	
(11)	$A \cup \emptyset = A$	
(12)	$A \cap U = A$	
(13)	$A \cup U = U$	
(14)	$A \cap \emptyset = \emptyset$	
(15)	$A \cup \overline{A} = U$	
(16)	$A \cap \overline{A} = \emptyset$	
(17)	$\overline{\overline{A}} = A$	
(18)	$\overline{U} = \emptyset$	
(19)	$\overline{\emptyset} = U$	
(20)	$A - B \subseteq A$	
(21)	If $A \subseteq B$ and $C \subseteq D$ then $(A \cup C) \subseteq (B \cup D)$	
(22)	If $A \subseteq B$ and $C \subseteq D$ then $(A \cap C) \subseteq (B \cap D)$	
(23)	$A \subseteq A \cup B$	
(24)	$A \cap B \subseteq A$	
(25)	If $A \subseteq B$ then $A \cup B = B$	
(26)	If $A \subseteq B$ then $A \cap B = A$	
(27)	$A - \emptyset = A$	
(28)	$A \cap (B - A) = \emptyset$	
(29)	$A \cup (B - A) = A \cup B$	
(30)	$A - (B \cup C) = (A - B) \cap (A - C)$	
(31)	$A - (B \cap C) = (A - B) \cup (A - C)$	

Another important concept in set theory is to look at the subsets of a given set. This leads to the definition of the power set.

Definition 5.8 (Power Set). The set of all subsets of a set A is the *power set of A*, denoted as $\wp(A)$.

If a set is finite, the power set is finite; if a set is infinite, the power set is infinite. The power set of a set with n elements has 2^n elements.

Example 5.6 The power set of $A = \{1, 2, 3\}$ with three elements has $2^3 =$ 8 elements and is written as $\wp(A) = \{\emptyset, \{1\}, \{2\}, \{3\}, \{1, 2\}, \{1, 3\}, \{2, 3\}, \{1, 2, 3\}\}$. Note that the empty set and the set itself are always elements of the power set.

5.4   Applications in GIS

Overlay operations are among the most common functions that a GIS provides for spatial analysis. Since spatial features such as points, arcs and polygons can be regarded as sets, overlay operations correspond to set intersection, union, difference, and complement. Table 5.2 shows the tools of the ArcGIS Pro overlay toolset and the corresponding set operations in mathematical notation.

Table 5.2   ArcGIS Pro overlay commands

Command	A	B	Set Operation
Erase	in_features	erase_features	$A - B$
Identity	in_features	identity_features	$A \cup (A \cap B)$
Intersect	in_features		$\bigcap_{i=1}^{n} A_i$
Symmetrical difference	in_features	update_features	$(A - B) \cup (B - A)$
Union	in_features		$\bigcup_{i=1}^{n} A_i$
Update	in_features	update_features	$(A - B) \cup B$

Normally, it does not matter in which sequence we apply overlay operations of the same type. The associative and commutative laws for set operations allow the application of intersection and union in arbitrary order.

The distributive laws can be used to simplify spatial overlay operations by reducing the number of operations. For example, if we have three data sets A, B and C. We need the intersection of A and B and the intersection of A and C, and finally, compute the union of the results. These operations amount to the following set operations $(A \cap B) \cup (A \cap C)$. This would need three overlay operations. However, the distributive law of set operations allows us to reduce the number of operations to two as $A \cap (B \cup C)$.

When we deal with polygon features in a GIS, we always have an embedding polygon that contains all features of our data set. Often it is called world polygon. In set theoretic terms, this corresponds to the universe of discourse.[4]

5.5 Exercises

Exercise 5.1 Highlight in the following Venn diagram $\overline{A} \cap \overline{B}$.

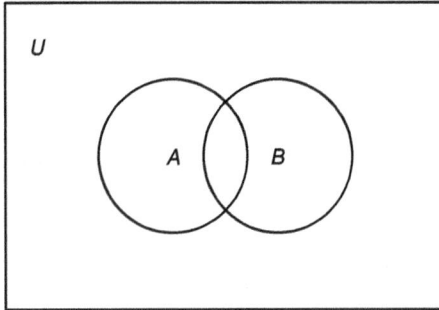

Exercise 5.2 Highlight in the following Venn diagram $A \cap (B \cup C)$.

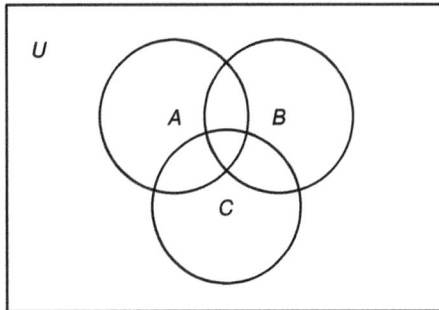

---

[4]In Section 9.5.1, we will see that also for topological reasons we need an embedding space for the cell complex of spatial features.

Exercise 5.3   Highlight in the following Venn diagram $\overline{A \cup B}$.

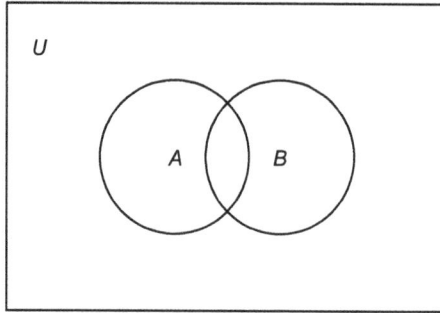

Exercise 5.4   Specify the power set for each of the following sets:
(1) $\{a, b, c\}$.
(2) $\{\{a, b\}, \{c\}\}$.
(3) $\{\emptyset\}$

Exercise 5.5   Let $U = \{a, b, c, d, e, f, g\}$ be the universe. $A = \{a, b, c, d, e\}$, $B = \{a, c, e, g\}$ and $C = \{b, e, f, g\}$ are sets. Compute the following:
(1) $\overline{B} \cup C$.
(2) $\overline{C} \cap A$.
(3) $B - C$.
(4) The power set of $B - C$.

Chapter 6

Relations and Functions

Relations are a very important concept in mathematics. Based on the fundamental principle of the Cartesian product we will introduce relations as the foundation of mappings and functions. Relations are based on a common understanding of relationships among objects. These relationships may refer to a comparison between objects of the same set, or they involve elements of different sets. Two special types of relations, the equivalence relation and the order relation, play an important role in mathematics. The first one is used to classify objects; the latter one is the basis for the theory of ordered sets. In this chapter, we deal only with binary relations, relations between two sets.

6.1 Cartesian Product

Definition 6.1 (Cartesian Product). The *Cartesian product* (or *cross product*) of two sets A and B, denoted as $A \times B$, is the set of all pairs $\{\langle a, b \rangle | a \in A \wedge b \in B\}$.

Example 6.1   Let $A = \{1, 2\}$, $B = \{a, b\}$ and $C = \emptyset$. Then
- $A \times B = \{\langle 1, a \rangle, \langle 1, b \rangle, \langle 2, a \rangle, \langle 2, b \rangle\}$.
- $B \times A = \{\langle a, 1 \rangle, \langle a, 2 \rangle, \langle b, 1 \rangle, \langle b, 2 \rangle\}$.
- $A \times C = \emptyset$.

Example 6.2   Consider the sets $A=\{$Vienna, Amsterdam$\}$ and $B=\{$Austria, Netherlands, France$\}$. The Cartesian product $A \times B$ is the set of six elements $\{\langle$Vienna, Austria$\rangle$, $\langle$Vienna, Netherlands$\rangle$, $\langle$Vienna, France$\rangle$, $\langle$Amsterdam, Austria$\rangle$, $\langle$Amsterdam, Netherlands$\rangle$, $\langle$Amsterdam, France$\rangle\}$.

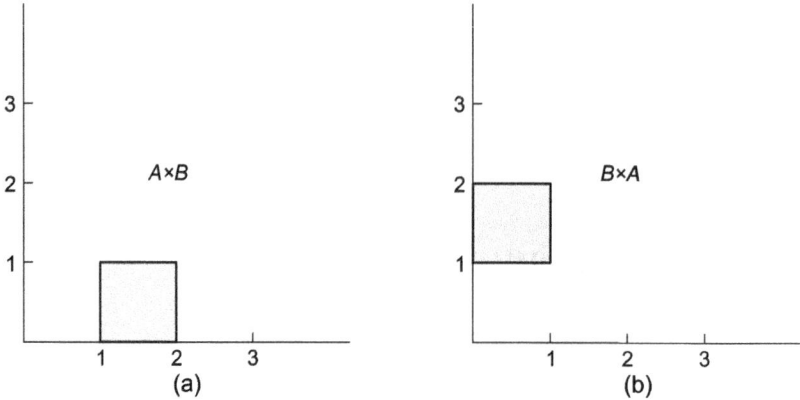

Figure 6.1   Non-commutativity of the Cartesian product

The Cartesian product is not commutative, i.e., $A \times B \neq B \times A$. This can easily be seen in Example 6.1.

We can also represent the Cartesian product graphically. Assume we have two sets $A = \{x | 1 \leqslant x \leqslant 2\}$ and $B = \{y | 0 \leqslant y \leqslant 1\}$. The Cartesian products of both sets $A \times B = \{\langle x, y \rangle | 1 \leqslant x \leqslant 2 \land 0 \leqslant y \leqslant 1\}$ and $B \times A = \{\langle y, x \rangle | 1 \leqslant x \leqslant 2 \land 0 \leqslant y \leqslant 1\}$ can be graphically represented as in Figure 6.1.

Some properties of the Cartesian product are listed in Table 6.1.

Table 6.1   Properties of the Cartesian product

(1) $A \times (B \cup C) = (A \times B) \cup (A \times C)$
(2) $A \times (B \cap C) = (A \times B) \cap (A \times C)$
(3) $(A \cup B) \times C = (A \times C) \cup (B \times C)$
(4) $(A \cap B) \times C = (A \times C) \cap (B \times C)$

6.2   Binary Relations

Although relations are generally defined with more than two sets, we restrict ourselves here to binary relations between two sets.

Definition 6.2 (Binary Relation). A *binary relation R over* $A \times B$ is a subset of $A \times B$. The set A is called the *domain* of R; B is the *codomain*. We write $\langle a, b \rangle \in R$ also as $a \sim b$, and $\langle a, b \rangle \notin R$ is written as $a \nsim b$. If the relation is defined over $A \times A$, we call it a relation on A.

Example 6.3   Consider the set $A = \{$Vienna, Amsterdam$\}$ and the set $B = \{$Austria, Netherlands, France$\}$. The set $C = \{\langle$Vienna, Austria$\rangle, \langle$Amsterdam, Netherlands$\rangle\}$ is a relation over $A \times B$ that can be read as "is the capital of".

Definition 6.3 (Inverse Relation).   Let R be a relation over $A \times B$. The *inverse relation* (or *inverse*) R^{-1} is defined as the relation over $B \times A$ such that $R^{-1} = \{\langle b, a \rangle | \langle a, b \rangle \in R\}$.

Example 6.4   Let $A = \{$John, Ann, Frank$\}$ and $B = \{$Mercedes, BMW$\}$ be two sets of persons and cars. $R = \{\langle$John, Mercedes$\rangle, \langle$Ann, BMW$\rangle, \langle$Frank, BMW$\rangle\}$ is a relation "drives a". Then $R^{-1} = \{\langle$Mercedes, John$\rangle, \langle$BMW, Ann$\rangle, \langle$BMW, Frank$\rangle\}$ is the inverse relation that can be read as "is driven by".

6.2.1   Relations and Predicates

Every binary relation R on a set A corresponds to a predicate with two variables and A as the universe of discourse. If the relation is given, the predicate can be defined as $P(a_1, a_2)$ is true if and only if $\langle a_1, a_2 \rangle \in R$. Likewise, if a predicate P is given we can define a relation R such that $R = \{\langle a_1, a_2 \rangle | P(a_1, a_2)$ is true$\}$.

6.2.2   Graphic Representation of Binary Relations

It is often convenient to represent relations graphically. For this purpose, we will use *directed graphs*[1] (or *digraphs*). If there is a relation between the two elements x and y, i.e., $x \sim y$, we use the digraph representation in Figure 6.2.

$$x \bullet \longrightarrow \bullet y$$

Figure 6.2   Binary relation represented by a digraph

---

[1]A graph is defined by two sets V and E, the set of *nodes* (*points* or *vertices*) and the set of *edges* (*arcs* or *lines*), and an incidence relation on $V \times V$ that describes which nodes are connected by edges. If the arcs are directed, we call the graph a *directed graph*.

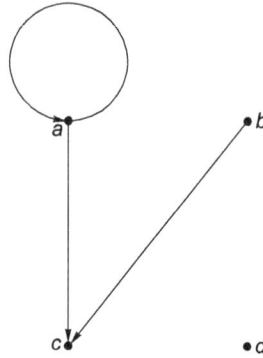

Figure 6.3   Relation R represented by a digraph

Example 6.5   Let $A = \{a, b, c, d\}$ be a set and $R = \{\langle a, c \rangle, \langle a, a \rangle, \langle b, c \rangle\}$ be a relation on A. The digraph is represented by the diagram of Figure 6.3.

6.2.3   Special Properties of Relations

Some properties of binary relations are so important that they must be discussed in more detail. The following list defines these properties.

Definition 6.4 (Properties of Relations).   Let R be a binary relation on a set A. We say that

- R is *reflexive* if $x \sim x$ for every x in A.
- R is *irreflexive* if $x \sim x$ for no x in A.
- R is *symmetric* if $x \sim y$ implies $y \sim x$ for every x, y in A.
- R is *antisymmetric* if $x \sim y$ and $y \sim x$ together imply $x = y$ for every x, y in A.
- R is *transitive* if $x \sim y$ and $y \sim z$ together imply $x \sim z$ for every x, y, z in A.

These properties are reflected in certain characteristics of a digraph representation of relations. The digraph of a reflexive relation has a loop on every node of the graph. The graph of an irreflexive relation has no loop on any node. A relation can be neither reflexive nor irreflexive. In this case, it simply has loops on some nodes, but not on all.

The graph of a symmetric relation has either two or no arcs between any two distinct nodes of the graph. For an antisymmetric relation the graph has

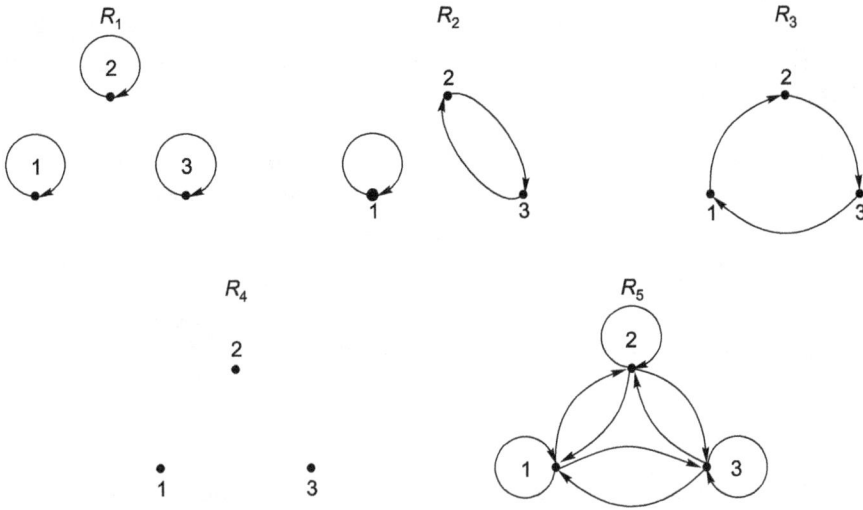

Figure 6.4   Sample relations

either one arc or no arc between any two distinct nodes of the graph. Loops may, but need not, occur in the graphs of symmetric and antisymmetric relations.

If in the graph of a transitive relation, there are two arcs from x to y and from y to z, then there must also be an arc from x to z.

Example 6.6   Consider the set of three elements $\{1, 2, 3\}$ and the relations represented in Figure 6.4.

- R_1 is reflexive, symmetric, antisymmetric, and transitive. It is the equality relation on the set. It is not irreflexive.
- R_2 is symmetric but not reflexive, irreflexive, antisymmetric, or transitive.
- R_3 is irreflexive and antisymmetric, but not reflexive, symmetric or transitive.
- R_4 is irreflexive, symmetric, antisymmetric and transitive. It is not reflexive. It is the empty relation on the set.
- R_5 is the universal relation on the set. It is reflexive, symmetric and transitive, but not irreflexive or antisymmetric.

6.2.3.1   *Equivalence Relation*

A reflexive, symmetric and transitive relation is called an *equivalence relation*. An equivalence relation divides a set S into nonempty mutually disjoint sets or *equivalence classes* $[a] = \{x | \langle a, x \rangle \in R\}$, where a is an element of S and

R is an equivalence relation. The set of all equivalence classes of S (written as S/R) is called the *quotient set* of S under R, i.e., $S/R = \{[a]|a \in R\}$. An element $y \in [a]$ is called a *representative* of the class $[a]$.

Example 6.7 Let R be the relation $\|$ (parallel) on the set of all lines in the plane. This relation is an equivalence relation, because (i) for every line l we have $l\|l$ (reflexive), (ii) for every two lines l_1, l_2 we have that if $l_1\|l_2$ then $l_2\|l_1$ (symmetric), and (iii) if $l_1\|l_2$ and $l_2\|l_3$ then $l_1\|l_3$ (transitive). The relation classifies the set of lines into the equivalence classes of parallel lines. Every element of one class is a representative of this class.

6.2.3.2   Order Relation

A reflexive, antisymmetric and transitive relation is called *order relation*. Order relations allow the comparison of elements of a set.

Example 6.8 The subset relation between two sets is an order relation, because (i) for all sets we have $A \subseteq A$ (reflexive), (ii) if $A \subseteq B$ and $B \subseteq A$ then it follows $A = B$ (antisymmetric), and (iii) if $A \subseteq B$ and $B \subseteq C$ then $A \subseteq C$ (transitive).

6.2.4   Composition of Relations

We can generate new relations by composing a sequence of relations. Formally, we define the composition of relations as Definition 6.5.

Definition 6.5 (Composition of Relations). Let R_1 be a relation from A to B, and R_2 be a relation from B to C. The composite relation from A to C, written as $R_1 R_2$ is defined as

$$R_1 R_2 = \{\langle a, c\rangle | a \in A \wedge c \in C \wedge \exists b[b \in B \wedge \langle a, b\rangle \in R_1 \wedge \langle b, c\rangle \in R_2]\}.$$

The composition of relations is not commutative, but associative.

A relation R on a set A can be composed with itself any number of times to form a new relation on the set A. For RR we also write R^2, for RRR we write R^3, and so on.

Example 6.9 If R is the relation "is father of", then RR is the relation "is grandfather of". Let $A = \{a, b, c, d\}$ be a set and consider $R_1 = \{\langle a, a\rangle, \langle a, b\rangle,$

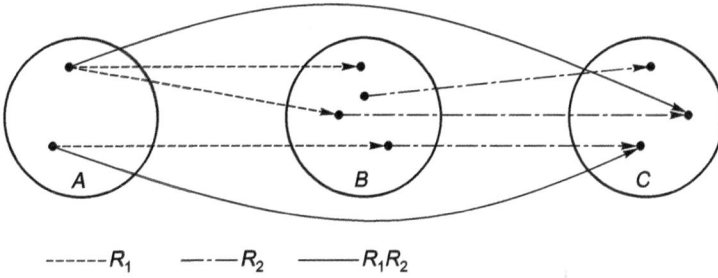

Figure 6.5   Composition of relations

$\langle b, d \rangle\}$ and $R_2 = \{\langle a, d \rangle, \langle b, c \rangle, \langle b, d \rangle, \langle c, b \rangle\}$ to be two relations on A. Then $R_1 R_2 = \{\langle a, c \rangle, \langle a, d \rangle\}$, $R_2 R_1 = \{\langle c, d \rangle\}$, $R_1^2 = \{\langle a, a \rangle, \langle a, b \rangle, \langle a, d \rangle\}$, and $R_2^3 = \{\langle b, c \rangle, \langle c, b \rangle, \langle b, d \rangle\}$.

The composition of relations can be illustrated with a digraph as displayed in Figure 6.5.

Let R_1 be a relation from A to B, R_2 and R_3 be two relations from B to C, and R_4 be a relation from C to D. Then the following statements are true:

(1) $R_1(R_2 \cup R_3) = R_1 R_2 \cup R_1 R_3$.
(2) $R_1(R_2 \cap R_3) \subseteq R_1 R_2 \cap R_1 R_3$.
(3) $(R_2 \cup R_3)R_4 = R_2 R_4 \cup R_3 R_4$.
(4) $(R_2 \cap R_3)R_4 \subseteq R_2 R_4 \cap R_3 R_4$.

6.3   Functions

Functions are a special kind of binary relations. They are used throughout mathematics.

Definition 6.6 (Function). A *function* (*map, mapping or transformation*) f from A to B, written as $f : A \to B$, is a binary relation from A to B such that for every $a \in A$, there exists a unique $b \in B$ such that $\langle a, b \rangle \in f$. We write $f(a) = b$ and we call A the *domain* and B the *codomain* of f; a is the *argument* and b is the *value* of the function for the argument a.

To correctly specify a function, we must indicate the domain, codomain and the value $f(x)$ for every argument x. Note that the important difference between a relation and a function is that for a function it is not possible that an

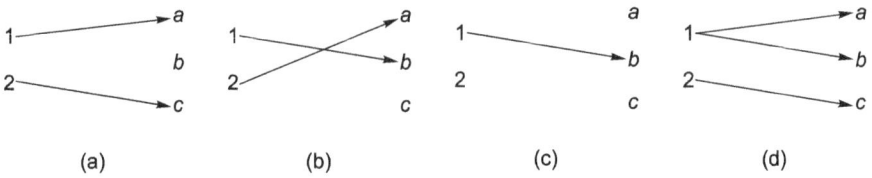

Figure 6.6 Functions and relations

argument has more than one value, and a value must exist for every element of the domain.

Example 6.10   Consider the function from the natural numbers to the natural numbers $f : \mathbb{N} \rightarrow \mathbb{N}$, where $f(x) = 2x - 1$. This function maps all natural numbers to the odd numbers. One is mapped to one, two to three, three to five, etc.

Example 6.11   Consider the sets $A = \{1, 2\}$ and $B = \{a, b, c\}$. When the domain and codomain are finite, we can represent functions as digraphs. In Figure 6.6 (a) and (b) are functions; (c) and (d) are not functions. (c) is not a function because not for every element of the domain we have a value. (d) is not a function because the argument 1 has more than one value.

6.3.1   Composition of Functions

In the same way as for relations, we can generate new functions by composing a sequence of functions.

Definition 6.7 (Composition of Functions). Let $g : A \rightarrow B$ and $f : B \rightarrow C$ be two functions. The composite function $f \circ g$ is a function from A to C and $(f \circ g)(x) = f(g(x))$ for all x in A. The composition of functions is not commutative, but it is associative.

Note that a composite function is only defined when the codomain of the first function g is equal to the domain of the second function f.

Example 6.12   Let $g : \mathbb{N} \rightarrow \mathbb{N}$ with $g(x) = 2x$ and $f : \mathbb{N} \rightarrow \mathbb{N}$ with $f(x) = x + 1$. The composite function $f(g(x)) = 2x + 1$ and the composite function $g(f(x)) = 2x + 2$.

6.3.2 Classes of Functions

Certain characteristics of functions are so important that a special terminology has been developed for them.

Definition 6.8 (Surjection). A function f from A to B is called *surjective* (*onto* or *surjection*) if the image of the domain is the codomain, or $f(A) = B$.

Definition 6.9 (Injection). A function f from A to B is called *injective* (*one-to-one* or *injection*) if distinct arguments have distinct values, or if $a \neq a'$ then $f(a) \neq f(a')$.

Definition 6.10 (Bijection). A function f from A to B is *bijective* (*one-to-one and onto*, or *bijection*) if it is surjective and injective.

Example 6.13 Let $f : \mathbb{Z} \to \{0, 1\}$ be a function from the integers to the set $\{0, 1\}$ defined by $f(x) = \begin{cases} 0 & \text{for } x \text{ is even} \\ 1 & \text{for } x \text{ is odd} \end{cases}$. This function is surjective, but not injective.

Example 6.14 Consider the function $f : \mathbb{Z} \to \mathbb{Z}$ in the integers with $f(x) = 2x - 1$. This function is injective, but not surjective.

Example 6.15 Consider the function $f : \mathbb{Z} \to \mathbb{Z}$ in the integers with $f(x) = x + 1$. This function is bijective.

In the case of functions from the real numbers to the real numbers, we can interpret the properties of being surjective, injective or bijective in terms of the graphs of the functions:
- **Surjectivity**: Every horizontal line intersects the graph of the function at least once.
- **Injectivity**: No horizontal line intersects the graph of the function more than once.
- **Bijectivity**: Every horizontal line intersects the graph of the function exactly once.

Example 6.16 Consider the function $f : \mathbb{R} \to \mathbb{R}$ with $f(x) = x^3 + 2x^2$. Every horizontal line intersects the graph of the function at least once. Therefore, the function is surjective. The function is not injective, because there are lines (e.g., $y = 0$) that intersect the graph more than once as shown in Figure 6.7.

Example 6.17   Consider the function $f : \mathbb{R} \to \mathbb{R}$ with $f(x) = 2^x + 10$. No horizontal line intersects the graph more than once. Therefore, the function is injective. It is not surjective, because there are lines that do not intersect the graph at all, as shown in Figure 6.8.

Example 6.18   Consider the function $f : \mathbb{R} \to \mathbb{R}$ with $f(x) = x$. Every horizontal line intersects the graph of the function exactly once. Therefore, the function is bijective as shown in Figure 6.9.

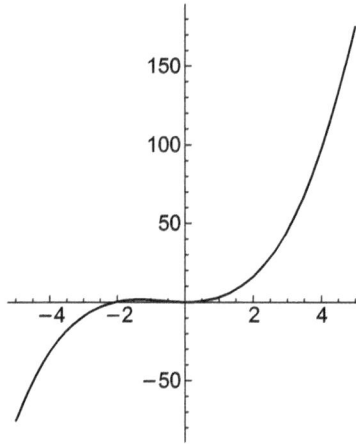

Figure 6.7   Example of a surjective but not injective function

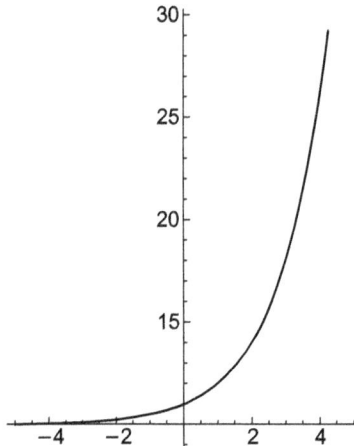

Figure 6.8   Example of an injective but not surjective function

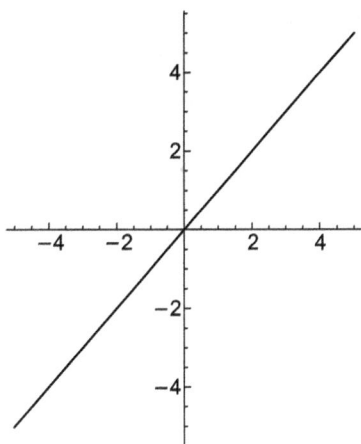

Figure 6.9   Example of a bijective function

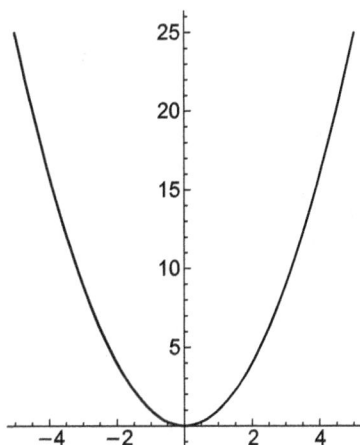

Figure 6.10   Example of a neither surjective nor injective function

Example 6.19   Consider the function $f : \mathbb{R} \to \mathbb{R}$ with $f(x) = x^2$. The function is neither surjective nor injective as shown in Figure 6.10.

These special properties of functions also propagate through composite functions. If $f \circ g$ is a composite function, then
- If f and g are surjective, then $f \circ g$ is surjective.
- If f and g are injective, then $f \circ g$ is injective.
- If f and g are bijective, then $f \circ g$ is bijective.

The converse of these statements is not true. However, we can establish the following:

- If $f \circ g$ is surjective, then f is surjective.
- If $f \circ g$ is injective, then g is injective.
- If $f \circ g$ is bijective, then f is surjective and g is injective.

Definition 6.11 (Inverse Function). Let $f : A \rightarrow B$ be a bijection from A to B. The converse relation of f is called the inverse function of f, written as f^{-1}.

The inverse function is only defined when the function is a bijection. The inverse function then is also a bijection.

6.4   Applications in GIS

Relations play an important role in GIS. The best-known examples of relations in GIS are the spatial or topological relations between the building blocks of feature data sets. These building blocks correspond to nodes, arcs and polygons in a two-dimensional setting.

Formally, we distinguish between the following relations among the elements of the set of nodes, arcs, and polygons. Every arc has a relation with two nodes (the start node and the end node relation); every arc has a relation with two polygons (the left polygon and the right polygon relation). Figure 6.11 shows a two-dimensional data set and the topological relations between nodes, arcs and polygons.

In this example we have the sets of nodes N, arcs A, and polygons P defined as $N = \{1, 2, 3, 4, 5, 6, 7\}$, $A = \{a, b, c, d, e, f, g, h, i, j\}$, and $P = \{A, B, C, D, E\}$. The start node—end node relation is defined as $AN = \{\langle a, 1 \rangle, \langle a, 2 \rangle, \langle b, 2 \rangle, \langle b, 3 \rangle, \langle c, 3 \rangle, \langle c, 1 \rangle, \langle d, 2 \rangle, \langle d, 4 \rangle, \langle e, 4 \rangle, \langle e, 6 \rangle, \langle f, 3 \rangle, \langle f, 6 \rangle, \langle g, 6 \rangle, \langle g, 5 \rangle, \langle h, 4 \rangle, \langle h, 5 \rangle, \langle i, 7 \rangle, \langle j, 5 \rangle, \langle j, 1 \rangle\}$, whereas the left polygon—right polygon relation can be written as $AP = \{\langle a, C \rangle, \langle a, 0 \rangle, \langle b, B \rangle, \langle b, 0 \rangle, \langle c, A \rangle, \langle c, 0 \rangle, \langle d, C \rangle, \langle d, B \rangle, \langle e, D \rangle, \langle e, B \rangle, \langle f, B \rangle, \langle f, A \rangle, \langle g, D \rangle, \langle g, A \rangle, \langle h, C \rangle, \langle h, D \rangle, \langle i, D \rangle, \langle i, E \rangle, \langle j, C \rangle, \langle j, A \rangle\}$.

Other types of relations are those among spatial features in a data set. The best-known examples are the eight relations between simple spatial regions that can be derived from topological invariants of boundary and interior (see Chapter 9). Figure 6.12 shows these relations.

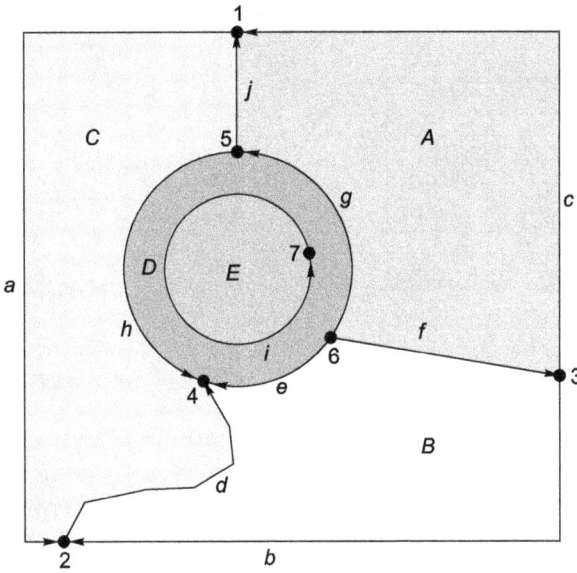

Figure 6.11 Topological relations. The start node and the end node of *i* are both node 7. The outside is regarded as a special polygon (indicated as 0) such that arcs *a*, *b* and *c* have *a* relation with two polygons

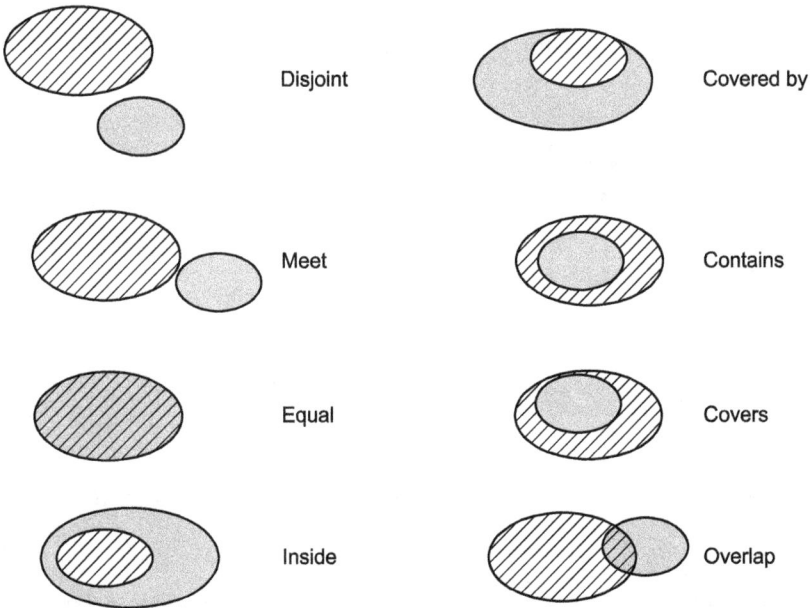

Figure 6.12 Spatial relations derived from topological invariants

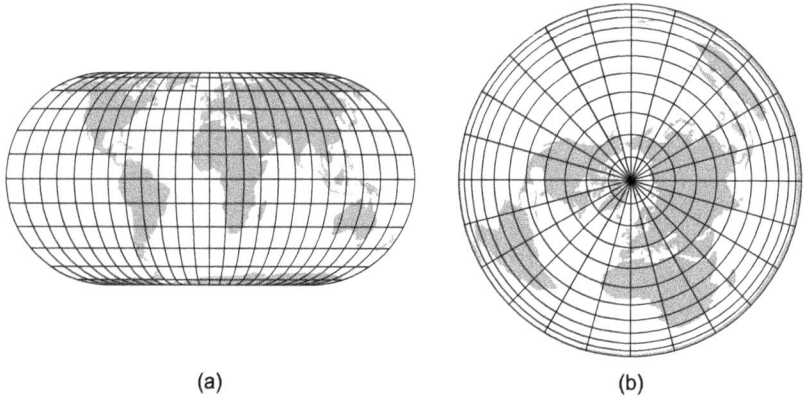

(a)   (b)

Figure 6.13   Map projections with singularities. (a) Eckert IV and (b) Azimuthal equal area

Functions appear in many different forms in GIS. One typical application is map projections. Here, a point on the surface of the earth whose location is given as latitude ϕ and longitude λ is mapped to a point on a plane by a set of two mapping rules for the easting and northing, respectively, as

$$\text{easting} = f_1(\phi, \lambda)$$
$$\text{northing} = f_2(\phi, \lambda)$$

Not every map projection is a function in the mathematical sense. Many map projections, for instance, map the pole to a line, which means that there is more than one value for a given argument. These cases, where a point on the earth is mapped to a line or cannot be mapped at all, are called singularities. Figure 6.13 shows two projections where the poles are mapped to a line (Figure 6.13a) and to a point (Figure 6.13b). In the second projection, the South Pole cannot be mapped at all.

6.5   Exercises

Exercise 6.1   Let $A = \{a, b\}$, $B = \{2, 3\}$ and $C = \{3, 4\}$. Compute:
(1) $A \times (B \cup C)$.
(2) $(A \times B) \cup (A \times C)$.
(3) $A \times (B \cap C)$.
(4) $(A \times B) \cap (A \times C)$.

Exercise 6.2   Let $W = \{1, 2, 3, 4\}$ and consider the following relations on W:

(1) $R_1 = \{\langle 1, 1 \rangle, \langle 1, 2 \rangle\}$.

(2) $R_2 = \{\langle 1, 1 \rangle, \langle 2, 3 \rangle, \langle 4, 1 \rangle\}$.

(3) $R_3 = \{\langle 1, 3 \rangle, \langle 2, 4 \rangle\}$.

(4) $R_4 = \{\langle 1, 1 \rangle, \langle 2, 2 \rangle, \langle 3, 3 \rangle\}$.

Check whether these relations are reflexive, irreflexive, symmetric, antisymmetric, or transitive.

Exercise 6.3   Let $X = \{1, 2, 3, 4\}$. Which one of the following relations is symmetric and which one is transitive? In case the relation is not symmetric or transitive, explain why.

(1) $f = \{\langle 2, 3 \rangle, \langle 1, 4 \rangle, \langle 2, 1 \rangle, \langle 3, 2 \rangle, \langle 4, 4 \rangle\}$.

(2) $g = \{\langle 3, 1 \rangle, \langle 4, 2 \rangle, \langle 1, 1 \rangle\}$.

(3) $h = \{\langle 2, 1 \rangle, \langle 3, 4 \rangle, \langle 1, 4 \rangle, \langle 2, 1 \rangle, \langle 4, 4 \rangle\}$.

Exercise 6.4   Let $X = \{1, 2, 3, 4\}$. Which one of the following relations is a function from X to X? In case the relation is not a function, explain why.

(1) $f = \{\langle 2, 3 \rangle, \langle 1, 4 \rangle, \langle 2, 1 \rangle, \langle 3, 2 \rangle, \langle 4, 4 \rangle\}$.

(2) $g = \{\langle 3, 1 \rangle, \langle 4, 2 \rangle, \langle 1, 1 \rangle\}$.

(3) $h = \{\langle 2, 1 \rangle, \langle 3, 4 \rangle, \langle 1, 4 \rangle, \langle 2, 1 \rangle, \langle 4, 4 \rangle\}$.

Exercise 6.5   Let R_1 and R_2 be relations on a set $A = \{a, b, c, d\}$ where $R_1 = \{\langle a, a \rangle, \langle a, c \rangle, \langle c, d \rangle\}$ and $R_2 = \{\langle a, d \rangle, \langle b, c \rangle, \langle b, b \rangle, \langle c, d \rangle\}$. Find $R_1 R_2$, $R_2 R_1$, R_1^2, and R_2^3.

Exercise 6.6   Let $f : \mathbb{R} \to \mathbb{R}$ be defined as $f(x) = x^2 - 3x + 2$. Compute $\dfrac{f(x + h) - f(x)}{h}$.

Exercise 6.7   Let f and g be functions on $X = \{1, 2, 3, 4, 5\}$. They are defined as:

$$f = \{\langle 1, 3 \rangle, \langle 2, 5 \rangle, \langle 3, 3 \rangle, \langle 4, 1 \rangle, \langle 5, 2 \rangle\}, \, g = \{\langle 1, 4 \rangle, \langle 2, 1 \rangle, \langle 3, 1 \rangle, \langle 4, 2 \rangle, \langle 5, 3 \rangle\}.$$

(1) Determine the codomain of f and g.

(2) Determine $g \circ f$ and $f \circ g$.

Chapter 7

Coordinate Systems and Transformations

All points in space can be uniquely referenced by their coordinates. Depending on the type of space, we distinguish between different coordinate systems such as Cartesian coordinate systems for Euclidean spaces, spherical coordinate systems for the sphere and elliptical coordinate systems for the ellipsoid. The sphere and the ellipsoid are geometric bodies used to approximate the shape of the earth.

Spatial features such as points, arcs and polygons as well as raster cells are spatially referenced through their coordinates. Often, it is necessary to apply transformations to these coordinates in order to shift, rotate, scale or warp the features. In this chapter, we will discuss the frequently used coordinate systems and transformations applied to geometric features in a Euclidean space.

7.1 Coordinate Systems

The coordinate system function is to assign any point in space a pair or triple of real numbers, namely its coordinates. The most common coordinate systems are rectangular or Cartesian coordinate systems and polar coordinate systems. In this chapter, we deal with a two- or three-dimensional real space (also called the *Euclidean space*) where every point has real-valued coordinates.

7.1.1 Cartesian Coordinate System

In the real plane $\mathbb{R}^2$ every point P has a unique pair of real numbers (x, y) assigned with $x, y \in \mathbb{R}$. On the other hand, every pair of real numbers (x, y) defines uniquely a point in the real plane. We define a single point O, the origin, and two perpendicular lines through that point, the axes. The horizontal axis is called the x-axis, and the vertical one is the y-axis. Every point P is uniquely defined by its *Cartesian coordinates* $P(x, y)$. Figure 7.1 illustrates the Cartesian coordinates of a point.

We can easily extend the two-dimensional coordinates in the plane to the three-dimensional coordinates in the space by defining a Cartesian coordinate system in $\mathbb{R}^2$. Every point P is then clearly defined by the triple (x, y, z) of Cartesian coordinates, as shown in Figure 7.2.

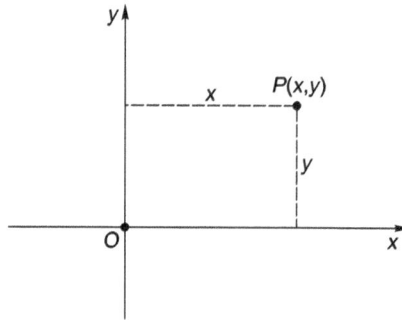

Figure 7.1   Cartesian coordinate system in the plane

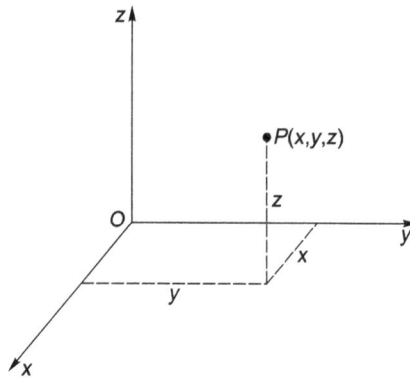

Figure 7.2   Cartesian coordinate system in three-dimensional space

7.1.2 Polar Coordinate System

A different way of assigning unique coordinates to a point in the plane is the use of polar coordinates. They are defined in a polar coordinate system which is given by a fixed point O, the *origin* or *pole*, and a line through the pole, the *polar axis*. Every point in the plane is then determined by its distance from the pole, the radius r, and the angle φ between the radius and the polar axis (Figure 7.3).

In a three-dimensional polar coordinate system (or *spherical coordinate system*) a point P is defined by the radius r from the origin to the point and two angles: the angle φ between the projection of $\overline{OP}$ onto the x, y-plane, and the angle θ between $\overline{OP}$ and the positive z-axis (Figure 7.4).

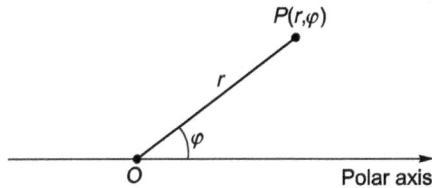

Figure 7.3   Polar coordinate system in the plane

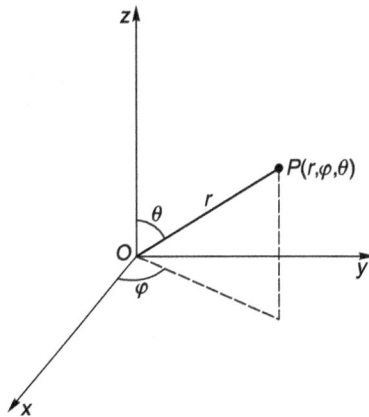

Figure 7.4   Polar coordinate system in three-dimensional space

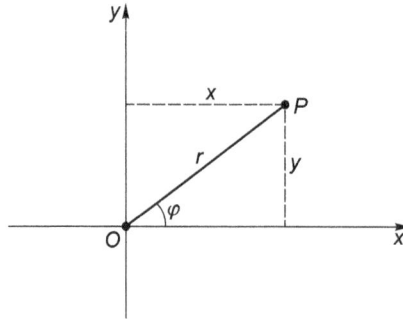

Figure 7.5   Conversion between Cartesian and polar coordinates in the plane

7.1.3   Transformations between Cartesian and Polar Coordinate Systems

The relationships between x and y, and r and φ are illustrated in Figure 7.5 and can be expressed by the following correspondences:

$$x = r \cos \varphi$$
$$y = r \sin \varphi$$
$$r = \sqrt{x^2 + y^2}$$
$$\tan \varphi = \frac{y}{x}, \varphi \in [0, 2\pi) \setminus \left\{ (2k + 1)\frac{\pi}{2} \middle| k \in \mathbb{Z} \right\}$$

Example 7.1   Given the Cartesian coordinates of the point $P(3, 4)$, we can compute its polar coordinates as $r = \sqrt{3^2 + 4^2} = \sqrt{9 + 16} = \sqrt{25} = 5$ and $\tan \varphi = \frac{4}{3} = 1.3333$, i.e., $\varphi = 53.13°$. The point thus has the polar coordinates $P(5, 53.13°)$.

The conversion between three-dimensional Cartesian coordinates and polar coordinates can be performed using the following formulas (see also Figure 7.6):

$$x = r \sin \theta \cos \varphi$$
$$y = r \sin \theta \sin \varphi$$
$$z = r \cos \theta$$
$$r = \sqrt{x^2 + y^2 + z^2}$$
$$\sin \varphi = \frac{y}{\sqrt{x^2 + y^2}}$$

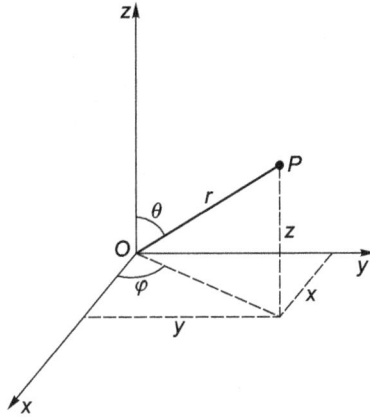

Figure 7.6 Conversion between Cartesian coordinates and polar coordinates in three-dimensional space

$$\cos\varphi = \frac{x}{\sqrt{x^2 + y^2}}$$

$$\cos\theta = \frac{z}{r}$$

$$\tan\theta = \frac{\sqrt{x^2 + y^2}}{z}, \theta \in [0, \pi] \setminus \left\{\frac{\pi}{2}\right\}$$

$$\tan\varphi = \frac{y}{x}, \varphi \in [0, 2\pi) \setminus \left\{(2k+1)\frac{\pi}{2} \middle| k \in \mathbb{Z}\right\}$$

Example 7.2 Given a point $P(2, 3, 4)$ in the three-dimensional Euclidean space $\mathbb{R}^3$ we can compute its polar coordinates as $r = \sqrt{2^2 + 3^2 + 4^2} = \sqrt{4 + 9 + 16} = \sqrt{29} = 5.385$, $\cos\theta = \frac{4}{5.385} = 0.743$ and $\tan\varphi = \frac{3}{2} = 1.5$. From this we get $\theta = 42.03°$ and $\varphi = 56.31°$.

7.1.4 Geographic Coordinate System

A special case of a spherical coordinate system is the *geographic coordinate system*, which is used to identify locations on the surface of the earth (Figure 7.7). The origin M of the geographic coordinate system is the center of the earth. The equator E lies in the plane defined by the x- and y-axes. The circle G defined by the intersection of the x, z-plane with the earth is the zero meridian through Greenwich. Every point P on the surface of the earth is uniquely defined by its latitude φ and longitude λ, denoted as $P(\varphi, \lambda)$.

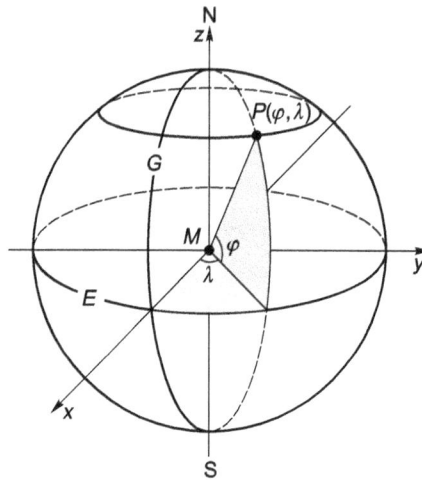

Figure 7.7   Geographic coordinate system

The latitude is measured as the angle between the equatorial plane and the radius from the origin to the point towards north (positive latitude) or south (negative latitude). A latitude on the northern hemisphere ranges from 0° to 90°, and from 0° to −90° on the southern hemisphere. Note that this is different from the way how the angle θ is defined for polar coordinates.

The longitude is the angle between the plane through the zero meridian and the origin, and the plane trough the meridian including the point P and the origin. Longitudes are positive from 0° to 180° towards east and negative from 0° to −180° towards west.

Every circle through the two poles is called a meridian; every circle parallel to the equatorial plane is called a parallel. For practical calculations, the radius R of the earth is assumed as 6,370 km.

Example 7.3   The airport of Vienna, Austria, has a latitude of 48°07′ North and a longitude of 16°34′ East, or VIE(48.1167°, 16.5667°).

7.2   Vectors and Matrices

Vectors and matrices play an important role in the analytical treatment of geometric figures. We can represent points in space by their respective point vectors, and we can apply many calculations related to the characteristics of geometric figures using vector representations.

7.2.1 Vectors

In Section 8.2.4, we will define the algebraic structure of a vector space. The elements of a vector space are called vectors. Several axioms for operations among vectors and vectors with scalars are defined. Here, we define vectors as a class of arrows in two- or three-dimensional real space.

Definition 7.1 (Vector). A *vector* is a class of parallel, directed arrows of the same length in space. A single arrow is called a *representative* of the vector.

For simplicity, we will not make a difference between a vector and a representative, and will simply call a representative a vector. The tail of a vector is called the *initial point*, and the head of a vector is called the *terminal point*.

Every point $P(x, y, z)$ in $\mathbb{R}^3$ can be represented by its point vector

$$P = \begin{pmatrix} x \\ y \\ z \end{pmatrix}$$

as shown in Figure 7.8.[1] In $\mathbb{R}^2$ the point vector components reduce to two for the x- and y-coordinate components.

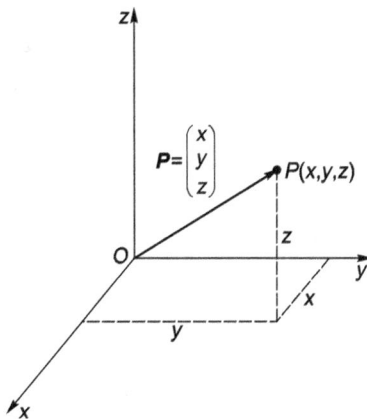

Figure 7.8 Point vector

---

[1]For the sake of a more compact vector notation, we also write $P = (x, y, z)$.

The *length* of a vector is defined as $|P| = \sqrt{x^2 + y^2 + z^2}$ in $\mathbb{R}^3$ and $|P| = \sqrt{x^2 + y^2}$ in $\mathbb{R}^2$. A vector of length 1 is called a *unit vector*.

In Section 8.2.4, we will see that we can define operations of addition and multiplication with a scalar for vectors:

$$a + b = \begin{pmatrix} a_x \\ a_y \\ a_z \end{pmatrix} + \begin{pmatrix} b_x \\ b_y \\ b_z \end{pmatrix} = \begin{pmatrix} a_x + b_x \\ a_y + b_y \\ a_z + b_z \end{pmatrix} \text{ and } \lambda a = \lambda \begin{pmatrix} a_x \\ a_y \\ a_z \end{pmatrix} = \begin{pmatrix} \lambda a_x \\ \lambda a_y \\ \lambda a_z \end{pmatrix}$$

Example 7.4   The sum of the two three-dimensional vectors $(1, 2, 3)$ and $(4, 5, 6)$ is

$$\begin{pmatrix} 1 \\ 2 \\ 3 \end{pmatrix} + \begin{pmatrix} 4 \\ 5 \\ 6 \end{pmatrix} = \begin{pmatrix} 1 + 4 \\ 2 + 5 \\ 3 + 6 \end{pmatrix} = \begin{pmatrix} 5 \\ 7 \\ 9 \end{pmatrix}$$

Beside the addition of vectors and the multiplication of a vector with a scalar, we know three different vector products. All of them also have geometric interpretations.

Definition 7.2 (Dot Product).   If a and b are two vectors, the *dot product* (or *Euclidean inner product*) is defined as

$$a \cdot b = \begin{pmatrix} a_x \\ a_y \\ a_z \end{pmatrix} \cdot \begin{pmatrix} b_x \\ b_y \\ b_z \end{pmatrix} = a_x b_x + a_y b_y + a_z b_z$$

The result of the dot product is always a number (scalar). The dot product is commutative and distributive, but not associative:

$$a \cdot b = b \cdot a$$
$$(a + b) \cdot c = a \cdot c + b \cdot c$$

The dot product can be used to compute the angle φ between two vectors a and b according to the following formula:

$$\cos \varphi = \frac{a \cdot b}{|a| \cdot |b|}$$

Example 7.5   The angle φ between the two vectors $a = (1, 2, 3)$ and $b = (3, 1, 1)$ is calculated according to the formula as

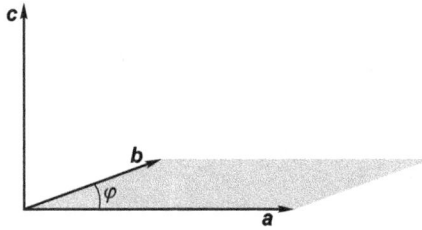

Figure 7.9   Cross product of two vectors

$$\cos \varphi = \frac{1 \cdot 3 + 2 \cdot 1 + 3 \cdot 1}{\sqrt{1+4+9} \cdot \sqrt{9+1+1}} = \frac{8}{\sqrt{14} \cdot \sqrt{11}} = 0.645.$$

It follows that $\varphi = 49.86°$.

Definition 7.3 (Cross Product). If a and b are two vectors, the cross product is defined as

$$c = a \times b = \begin{pmatrix} a_x \\ a_y \\ a_z \end{pmatrix} \times \begin{pmatrix} b_x \\ b_y \\ b_z \end{pmatrix} = \begin{pmatrix} a_y b_z - a_z b_y \\ a_z b_x - a_x b_z \\ a_x b_y - a_y b_x \end{pmatrix}$$

The result of the cross product is a vector. It is distributive, but not commutative:

$$(a + b) \times c = a \times c + b \times c$$

$$a \times b = -b \times a$$

The cross product can be geometrically interpreted as illustrated in Figure 7.9:
- The product vector c is perpendicular to the vectors a and b.
- a, b and c form a right-handed coordinate system.
- The length of c is equal to the area of the parallelogram spanned by a and b, where φ is the angle between the two vectors, according to the following formula $|c| = |a \times b| = |a| \cdot |b| \cdot \sin \varphi$.

Example 7.6   The cross product of the two vectors $a = (1, 0, 0)$ and $b = (1, 1, 0)$ is equal to

$$a \times b = \begin{pmatrix} 1 \\ 0 \\ 0 \end{pmatrix} \times \begin{pmatrix} 1 \\ 1 \\ 0 \end{pmatrix} = \begin{pmatrix} 0 \cdot 0 - 0 \cdot 1 \\ 0 \cdot 1 - 1 \cdot 0 \\ 1 \cdot 1 - 0 \cdot 1 \end{pmatrix} = \begin{pmatrix} 0 \\ 0 \\ 1 \end{pmatrix}$$

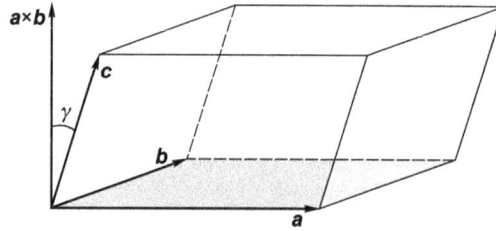

Figure 7.10   Scalar triple product

and the length of the vector is 1. The angle between a and b is 45°. Therefore, we can compute the length of the cross product as $1 \cdot \sqrt{2} \cdot \dfrac{\sqrt{2}}{2} = 1$.

Definition 7.4 (Scalar Triple Product). If a, b and c are three vectors, the scalar triple product is defined as

$$(abc) = (a \times b) \cdot c = c \cdot (a \times b)$$
$$= a_x(b_y c_z - b_z c_y) - a_y(b_x c_z - b_z c_x) + a_z(b_x c_y - b_y c_x)$$

If the three vectors do not lie in the same plane, then they form a parallelepiped when they are positioned with the common initial point (Figure 7.10). The result of the scalar triple product is a number equal to the volume of this parallelepiped and can also be computed according to the formula

$$(abc) = |a \times b| \cdot |c| \cdot \cos \gamma$$

where γ is the angle between the cross product of $a \times b$ and c.

Example 7.7   The volume of the parallelepiped with $a = (2, -6, 2)$, $b = (0, 4, -2)$, and $c = (2, 2, -4)$ is computed by inserting into the formula given in Definition 7.4 as $2 \cdot [4 \cdot (-4) - (-2) \cdot 2] - (-6) \cdot [0 \cdot (-4) - (-2) \cdot 2] + 2 \cdot [0 \cdot 2 - 4 \cdot 2] = 2 \cdot (-12) - (-6) \cdot 4 + 2 \cdot (-8) = -24 + 24 - 16 = -16$.

The following correspondences hold between the dot product, cross product and scalar triple product:

$$a \times (b \times c) = (a \cdot c)b - (a \cdot b)c$$
$$(a \times b) \cdot (c \times d) = (a \cdot c)(b \cdot d) - (a \cdot d)(b \cdot c)$$
$$(a \times b)^2 = a^2 b^2 - (a \cdot b)^2$$
$$(a \times b) \times (c \times d) = c(abd) - d(abc)$$

7.2.2 Matrices

Rectangular arrays of real numbers occur in many contexts in mathematics and as data structure in applications of computer science.

Definition 7.5 (Matrix). A *matrix* is a rectangular array of real numbers. The numbers in the array are called the entries in the matrix.

A matrix M is a rectangular array of numbers with m rows and n columns represented as:

$$M = \begin{pmatrix} a_{11} & a_{12} & a_{13} & \cdots & a_{1n} \\ a_{21} & a_{22} & a_{23} & \cdots & a_{2n} \\ \vdots & \vdots & \vdots & & \vdots \\ a_{m1} & a_{m2} & a_{m3} & \cdots & a_{mn} \end{pmatrix}$$

We call M a $m \times n$-matrix. We also know that the matrices form a vector space and that we can multiply matrices with matrices and matrices with vectors according to the following rules:

Let A and B be two matrices. Their sum is defined as

$$\begin{pmatrix} a_{11} & a_{12} & \cdots & a_{1n} \\ a_{21} & a_{22} & \cdots & a_{2n} \\ \vdots & \vdots & & \vdots \\ a_{m1} & a_{m2} & \cdots & a_{mn} \end{pmatrix} + \begin{pmatrix} b_{11} & b_{12} & \cdots & b_{1n} \\ b_{21} & b_{22} & \cdots & b_{2n} \\ \vdots & \vdots & & \vdots \\ b_{m1} & b_{m2} & \cdots & b_{mn} \end{pmatrix}$$

$$= \begin{pmatrix} a_{11} + b_{11} & a_{12} + b_{12} & \cdots & a_{1n} + b_{1n} \\ a_{21} + b_{21} & a_{22} + b_{22} & \cdots & a_{2n} + b_{2n} \\ \vdots & \vdots & & \vdots \\ a_{m1} + b_{m1} & a_{m2} + b_{m2} & \cdots & a_{mn} + b_{mn} \end{pmatrix}$$

Note that the sum has the same number of rows and columns as the input matrices, and that both matrices must have the same number of rows and columns. If the number of rows and columns do not match, the sum is not defined.

The *multiplication of a matrix A with a scalar s* is defined as

$$sA = s \begin{pmatrix} a_{11} & a_{12} & \cdots & a_{1n} \\ a_{21} & a_{22} & \cdots & a_{2n} \\ \vdots & \vdots & & \vdots \\ a_{m1} & a_{m2} & \cdots & a_{mn} \end{pmatrix} = \begin{pmatrix} sa_{11} & sa_{12} & \cdots & sa_{1n} \\ sa_{21} & sa_{22} & \cdots & sa_{2n} \\ \vdots & \vdots & & \vdots \\ sa_{m1} & sa_{m2} & \cdots & sa_{mn} \end{pmatrix}$$

The product of two matrices A and B is only defined if the number of columns of A is equal to the number of rows of B. Given an $m \times p$-matrix A and a $p \times n$-matrix B the product of A and B is an $m \times n$-matrix where every element c_{ij} of the product $C = AB$ is calculated according to the following schema

$$\begin{pmatrix} a_{11} & \cdots & a_{1p} \\ \cdot & \cdots & \cdot \\ a_{i1} & \cdots & a_{ip} \\ \cdot & \cdots & \cdot \\ a_{m1} & \cdots & a_{mp} \end{pmatrix} \begin{pmatrix} b_{11} & \cdots & b_{1j} & \cdots & b_{1n} \\ \cdot & \cdots & \cdot & \cdots & \cdot \\ \cdot & \cdots & \cdot & \cdots & \cdot \\ \cdot & \cdots & \cdot & \cdots & \cdot \\ b_{p1} & \cdots & b_{pj} & \cdots & b_{pn} \end{pmatrix} = \begin{pmatrix} c_{11} & \cdots & & c_{1n} \\ \cdot & \cdots & & \cdot \\ \cdot & & c_{ij} & \cdot \\ \cdot & \cdots & & \cdot \\ c_{m1} & \cdots & & c_{mn} \end{pmatrix}$$

and

$$c_{ij} = a_{i1}b_{1j} + a_{i2}b_{2j} + \cdots + a_{ip}b_{pj} = \sum_{k=1}^{p} a_{ik}b_{kj}$$

Example 7.8 The product of the two matrices $\begin{pmatrix} r & s \\ t & u \end{pmatrix}$ and $\begin{pmatrix} a_1 & a_2 & a_3 \\ b_1 & b_2 & b_3 \end{pmatrix}$ is computed as

$$\begin{pmatrix} r & s \\ t & u \end{pmatrix} \begin{pmatrix} a_1 & a_2 & a_3 \\ b_1 & b_2 & b_3 \end{pmatrix} = \begin{pmatrix} ra_1 + sb_1 & ra_2 + sb_2 & ra_3 + sb_3 \\ ta_1 + ub_1 & ta_2 + ub_2 & ta_3 + ub_3 \end{pmatrix}$$

Example 7.9 The product of $\begin{pmatrix} 1 & 2 \\ 3 & 4 \end{pmatrix}$ and $\begin{pmatrix} 5 & 6 \\ 7 & 8 \end{pmatrix}$ is calculated as

$$\begin{pmatrix} 1 & 2 \\ 3 & 4 \end{pmatrix} \begin{pmatrix} 5 & 6 \\ 7 & 8 \end{pmatrix} = \begin{pmatrix} 1 \cdot 5 + 2 \cdot 7 & 1 \cdot 6 + 2 \cdot 8 \\ 3 \cdot 5 + 4 \cdot 7 & 3 \cdot 6 + 4 \cdot 8 \end{pmatrix} = \begin{pmatrix} 19 & 22 \\ 43 & 50 \end{pmatrix}$$

7.3 Transformations

When we rotate, shift or scale geometric figures we apply geometric transformations. Here, we will focus on plane coordinate systems and their transformations. Another problem related to transformations is to determine the parameters of a transformation between two plane coordinate systems that compensates for scaling, rotation, skew and translation. We will discuss the Helmert (or similarity) transformation and the affine transformation that both provide solutions to this problem.

7.3.1 Geometric Transformations

In the following sections, we will discuss geometric transformations in the plane using Cartesian coordinates.

7.3.1.1 *Translation*

The shift of a geometric figure in horizontal and vertical direction results in a translation operation (Figure 7.11). The translation factor in the x-direction is t_x, the factor in the y-direction is t_y. They need not be the same.

Given the coordinates of a point $P(x, y)$, the coordinates of the new point $P'(x', y')$ are calculated according to the following formula:

$$x' = x + t_x$$
$$y' = y + t_y$$

In matrix notation, we can write the translation of a point defined by its vector $\boldsymbol{P} = \begin{pmatrix} x \\ y \end{pmatrix}$ with a translation vector $\boldsymbol{t} = \begin{pmatrix} t_x \\ t_y \end{pmatrix}$ resulting in a new point $\boldsymbol{P'} = \begin{pmatrix} x' \\ y' \end{pmatrix}$ as $\boldsymbol{P'} = \boldsymbol{P} + \boldsymbol{t}$ or $\begin{pmatrix} x' \\ y' \end{pmatrix} = \begin{pmatrix} x \\ y \end{pmatrix} + \begin{pmatrix} t_x \\ t_y \end{pmatrix}$.

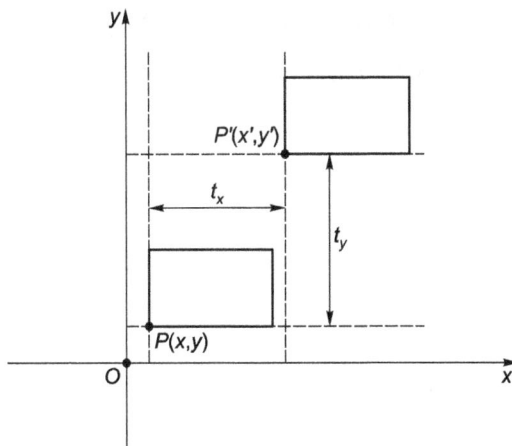

Figure 7.11 Translation

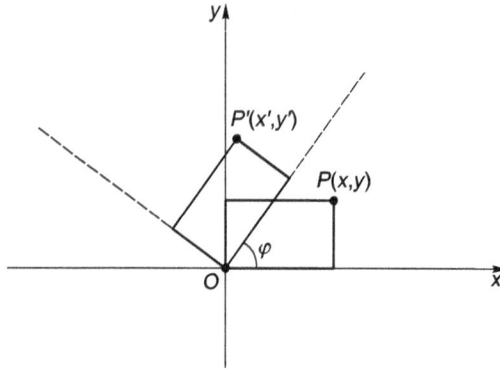

Figure 7.12   Rotation

7.3.1.2   Rotation

The rotation of a geometric figure in a two-dimensional coordinate system with an angle φ is shown in Figure 7.12.

Given the coordinates of a point $P(x, y)$, the coordinates of the rotated point $P'(x', y')$ are calculated according to the following formula:

$$x' = x \cos \varphi - y \sin \varphi$$
$$y' = x \sin \varphi + y \cos \varphi$$

In matrix notation, the rotation of a point $\boldsymbol{P} = \begin{pmatrix} x \\ y \end{pmatrix}$ with an angle φ can be denoted as $\boldsymbol{P}' = \boldsymbol{RP}$ with the rotation matrix

$$\boldsymbol{R} = \begin{pmatrix} \cos \varphi & -\sin \varphi \\ \sin \varphi & \cos \varphi \end{pmatrix}$$

or

$$\begin{pmatrix} x' \\ y' \end{pmatrix} = \begin{pmatrix} \cos \varphi & -\sin \varphi \\ \sin \varphi & \cos \varphi \end{pmatrix} \begin{pmatrix} x \\ y \end{pmatrix}$$

7.3.1.3   Scaling

The scaling of a geometric figure can be described by the application of a multiplication factor (or scaling factor) to the coordinates in a given coordinate system (Figure 7.13).

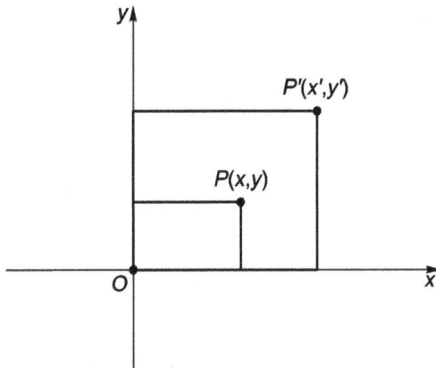

Figure 7.13 Scaling

Given the coordinates of a point $P(x, y)$, the coordinates of the scaled point $P'(x', y')$ are calculated according to the following formula:

$$x' = s_x \cdot x$$
$$y' = s_y \cdot y$$

The factors s_x and s_y are the multiplication factors in the x- and y-directions. They need not be the same. In matrix notation, the scaling of a point $P = \begin{pmatrix} x \\ y \end{pmatrix}$ with multiplication factors s_x and s_y for x and y, respectively, can be written as $P' = SP$ with the scaling matrix

$$S = \begin{pmatrix} s_x & 0 \\ 0 & s_y \end{pmatrix} \text{ or } \begin{pmatrix} x' \\ y' \end{pmatrix} = \begin{pmatrix} s_x & 0 \\ 0 & s_y \end{pmatrix} \begin{pmatrix} x \\ y \end{pmatrix}$$

7.3.2 Combination of Transformations

When we want to perform rotation, scaling and translation to a point, either we can apply the respective transformations in sequence, one after the other, or we can combine the transformation matrices to one matrix for rotation and scaling and add the translation vector. The general approach using the matrices defined in the previous sections is written as

$$P' = SRP + t$$

In the detailed notation this translates to

$$\begin{pmatrix} x' \\ y' \end{pmatrix} = \begin{pmatrix} s_x \cos\varphi & -s_x \sin\varphi \\ s_y \sin\varphi & s_y \cos\varphi \end{pmatrix} \begin{pmatrix} x \\ y \end{pmatrix} + \begin{pmatrix} t_x \\ t_y \end{pmatrix}$$

Example 7.10   Let S be a square defined with the points

$$P_1 = \begin{pmatrix} 0 \\ 0 \end{pmatrix}, \quad P_2 = \begin{pmatrix} 1 \\ 0 \end{pmatrix}, \quad P_3 = \begin{pmatrix} 1 \\ 1 \end{pmatrix}, \quad P_4 = \begin{pmatrix} 0 \\ 1 \end{pmatrix}$$

We apply a rotation of $45°$, a scaling with the factor 2 in x-direction and 1 in y-direction. Finally, we shift the figure three units to the right and two units up. The result can be computed according to the transformation matrix

$$T = \begin{pmatrix} \sqrt{2} & -\sqrt{2} \\ \dfrac{1}{2}\sqrt{2} & \dfrac{1}{2}\sqrt{2} \end{pmatrix}$$

and the translation vector

$$t = \begin{pmatrix} 3 \\ 2 \end{pmatrix} \text{ as } P' = TP + t$$

Figure 7.14 shows the original square and the transformed shape where

$$P_1' = \begin{pmatrix} 3 \\ 2 \end{pmatrix}, \quad P_2' = \begin{pmatrix} 3+\sqrt{2} \\ 2+\dfrac{1}{\sqrt{2}} \end{pmatrix}, \quad P_3' = \begin{pmatrix} 3 \\ 2+\sqrt{2} \end{pmatrix}, \quad P_4' = \begin{pmatrix} 3-\sqrt{2} \\ 2+\dfrac{1}{\sqrt{2}} \end{pmatrix}$$

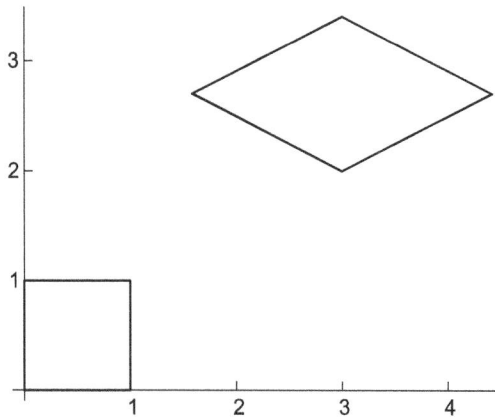

Figure 7.14   Example of combination of transformations

7.3.3 Homogeneous Coordinates

An easier way to deal with geometric transformations is to use homogeneous coordinates.

Definition 7.6 (Homogeneous Coordinates). Every point with the Cartesian coordinates (x, y) can be assigned the homogeneous coordinates $(t \cdot x, t \cdot y, t)$. Conversely, given the homogeneous coordinates of a point (r, s, t), we can determine its Cartesian coordinates as $\left(\dfrac{r}{t}, \dfrac{s}{t}\right)$.

We assign to a point $P(x, y)$ its homogeneous coordinates $(x, y, 1)$. The geometric transformations can then be expressed by 3×3-matrices.

$$R = \begin{pmatrix} \cos\varphi & -\sin\varphi & 0 \\ \sin\varphi & \cos\varphi & 0 \\ 0 & 0 & 1 \end{pmatrix} \quad \text{rotation}$$

$$S = \begin{pmatrix} s_x & 0 & 0 \\ 0 & s_y & 0 \\ 0 & 0 & 1 \end{pmatrix} \quad \text{scaling}$$

$$T = \begin{pmatrix} 1 & 0 & t_x \\ 0 & 1 & t_y \\ 0 & 0 & 1 \end{pmatrix} \quad \text{translation}$$

Note that now the translation can also be expressed through a translation matrix. This allows us to combine all three transformations in one single transformation matrix:

$$U = \begin{pmatrix} s_x \cos\varphi & -s_x \sin\varphi & t_x \\ s_y \sin\varphi & s_y \cos\varphi & t_y \\ 0 & 0 & 1 \end{pmatrix}$$

The general transformation of a point including rotation, scaling and translation can then be written simply as the multiplication of the transformation matrix with the point vector, namely $P' = UP$ or

$$\begin{pmatrix} x' \\ y' \\ 1 \end{pmatrix} = \begin{pmatrix} s_x \cos\varphi & -s_x \sin\varphi & t_x \\ s_y \sin\varphi & s_y \cos\varphi & t_y \\ 0 & 0 & 1 \end{pmatrix} \begin{pmatrix} x \\ y \\ 1 \end{pmatrix}$$

Example 7.11   The transformation of the square in the previous example can now be written as

$$\begin{pmatrix} x' \\ y' \\ 1 \end{pmatrix} = \begin{pmatrix} \sqrt{2} & -\sqrt{2} & 3 \\ \dfrac{\sqrt{2}}{2} & \dfrac{\sqrt{2}}{2} & 2 \\ 0 & 0 & 1 \end{pmatrix} \begin{pmatrix} x \\ y \\ 1 \end{pmatrix}$$

7.3.4   Transformation between Coordinate Systems

In many applications, we have to transform coordinates from one system into another coordinate system. In principle, this relates to the geometric transformations discussed in the previous sections. However, often we do not know the rotation angle, multiplication factor or translation vector, or there are more distortions involved. In these cases, we must determine the transformation parameters from known coordinates of points in both systems. These points are called *control points*. The most common transformations used are:

- similarity transformation;
- affine transformation;
- projective transformation.

The *similarity transformation* (also called *Helmert transformation*) scales, rotates, and translates the data. It does not independently scale the axes or introduce any skew. It is also known as four-parameter transformation and has the general form:

$$x' = Ax + By + C$$
$$y' = -Bx + Ay + F$$

A minimum of two control points is required to be able to compute the four parameters A, B, C, and F.

The *affine transformation* (or *six-parameter transformation*) will differentially scale, skew, rotate, and translate the data. It requires a minimum of three control points and has the general form:

$$x' = Ax + By + C$$
$$y' = Dx + Ey + F$$

Finally, the *projective transformation* (or *eight-parameter transformation*) can compensate for greater distortions between the two coordinate systems and requires a minimum of four control points. It has the general form:

$$x' = \frac{Ax + By + C}{Gx + Hy + 1}$$
$$y' = \frac{Dx + Ey + F}{Gx + Hy + 1}$$

Example 7.12 Given two control points P_1 and P_2, compute the parameters of a Helmert transformation. The coordinates of the control points are measured or given as $P_1(x_1, y_1)$ and $P_2(x_2, y_2)$. In the other coordinate system, the corresponding points are given as $P_1'(x_1', y_1')$ and $P_2'(x_2', y_2')$. We can now compute the parameters of the Helmert transformation by solving the following system of linear equations for A, B, C and F.

$$x_1' = Ax_1 + By_1 + C$$
$$y_1' = -Bx_1 + Ay_1 + F$$
$$x_2' = Ax_2 + By_2 + C$$
$$y_2' = -Bx_2 + Ay_2 + F$$

Normally, we use more than the required minimum number of control points. We then need to solve the resulting system of equations with a least squares approach. The *root mean square error* (RMSE) indicates the goodness of fit. Ideally, the best fit would result in an RMSE of zero, which is never the case when we use more than the required minimum number of control points. However, the RMSE should be as small as possible to achieve a reliable set of parameters for the transformation.

7.4 Applications in GIS

In GIS, we apply geometric transformations in many different ways. One use of transformations is in the graphic editing functions of every GIS. When we edit spatial features, we need to shift, rotate, skew and scale them.

Another important application lies in the transformation of coordinates of data sets as it occurs, for instance, when we have a map in an unknown map projection or in manual digitizing. Here, we have to set up a transformation

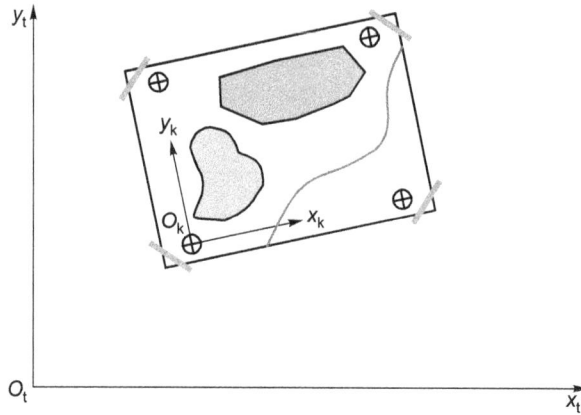

Figure 7.15   Manual digitizing setup

from the device coordinates, i.e., the coordinates produced by the digitizing device—usually in millimeters or inches—to the world coordinates, i.e., the coordinates of the map projection.

Figure 7.15 shows a sketch of a digitizing tablet with a map mounted on it. The origin of the tablet coordinates lies at O_t. The map coordinate system is indicated with O_k and the axes x_k and y_k. On the map, we have identified four ticks (or control points) designated with ⊕. The map coordinates of these ticks are either known or can be determined easily, for instance as grid intersection points or corner points of the map sheet whose coordinates can be read from the map.

The map is usually not aligned with the coordinate system of the tablet. Before we can start digitizing, we need to establish a relationship between the tablet coordinate and the map coordinate system. This is done by choosing a proper transformation and by computing its parameters. Usually, we select a four- or six-parameter transformation. With the given coordinates of the ticks (in the map coordinate system) and their measured coordinates (in the tablet coordinate system), we can compute the transformation parameters and subsequently apply the transformation to every point measured on the map. The transformation converts these coordinates into map coordinates.

Every GIS software package should provide this functionality for manual digitizing or general coordinate transformation. The Transform Features tool in ArcGIS Pro is one example of such a function.

7.5   Exercises

Exercise 7.1   Given the geographic coordinates of Vienna airport as $\varphi = 48.12°$, $\lambda = 16.57°$ and the radius of the Earth as $R = 6370$ km, compute the Cartesian coordinates of the airport when the origin of the Cartesian coordinate system is located at the center of the earth and the axis directions are as illustrated in Figure 7.7.

Exercise 7.2   Explain why each of the following expressions makes no sense when the operation "·" denotes the dot product of vectors:
(1) $a \cdot (b \cdot c)$.
(2) $a + (b \cdot c)$.
(3) $k \cdot (a \cdot b)$.

Exercise 7.3   Let $a = (1, 3, 2)$, $b = (1, 2, 3)$ and $c = (2, -1, 4)$. Compute (a) $b \times c$, (b) $(a \times b) - 2c$, and (c) $(a \times b) \times (b \times c)$.

Exercise 7.4   What is wrong with the expression $a \times b \times c$?

Exercise 7.5   Let $a = (1, 3, 2)$, $b = (1, 2, 3)$ and $c = (2, -1, 4)$. Compute (a, b, c).

Exercise 7.6   Given the triangle $\triangle ABC$ with the coordinates $A(1, 0)$, $B(3, 0)$ and $C(2, 1)$, compute the coordinates of the resulting figure when the triangle is rotated by 60°, scaled with a factor of 1.5 in both x- and y-direction, and translated 2 units to the right and 3 units down.

Exercise 7.7   Use a map sheet of your choice and perform the procedure for setting up the transformation parameters in manual digitizing using four control points and a six-parameter transformation.

Chapter 8

Algebraic Structures

Mathematical structures are used to describe real-world processes or phenomena. As we have seen before, there are three main structures in mathematics: algebraic structures, topologic structures, and order structures. In this chapter we will describe algebraic structures (or algebras) that are characterized by a set on which operations are defined for the elements of this set.

These operations are needed when we want to "compute" or "calculate" with the elements of a set. Often, we need to identify mappings from one algebra to another. If the structure is preserved under such a mapping, we call it a structure preserving mapping or homomorphism.

8.1 Components of An Algebra

Definition 8.1 (Algebra). Whenever we specify an algebra we need to describe the following components:
(1) A set S, the carrier of the algebra.
(2) Operations defined on the elements of the carrier.
(3) Distinguished elements of the carrier, the *constants* of the algebra.

The carrier S of an algebra is a set of elements on which operations are defined. Examples of carriers are number sets such as $\mathbb{N}$ (natural numbers), $\mathbb{Z}$ (integers), or $\mathbb{R}$ (real numbers). Operations are defined as a mapping $\circ : S^m \to S$, where the m is called the "arity" of the operation. Operations from $S^1 = S \to S$ are called *unary* operations. As an example of a unary operation consider the operation "$-$" that assigns the negative value to an element, i.e., it takes the number x to $-x$. *Binary* operations are mappings from $S^2 \to S$

and operate on two elements of the carrier. Examples of a binary operation are addition $x + y$ and multiplication $x \cdot y$ of elements. The constants of an algebra are distinguished elements of the carrier set with properties of special importance. Algebras are denoted as n-tuples ⟨carrier, operations, constants⟩.

Example 8.1   The real numbers $\mathbb{R}$ with the binary operations $+$ (addition) and $\cdot$ (multiplication), the unary operation $-$ and the constants 0 and 1 are an algebra that is represented as the six-tuples ⟨$\mathbb{R}, +, \cdot, -, 0, 1$⟩.

8.1.1   Signature and Variety

Often, we look at a class of algebras such that every member of the class has the same characteristics.

Definition 8.2 (Signature of An Algebra). Two algebras have the same *signature* (or are of the same *species*) if their n-tuples include the same number of operations and constants and the arities of corresponding operations are the same.

Example 8.2   The algebras ⟨$\mathbb{R}, +, \cdot, 1, 0$⟩ and ⟨$\wp(S), \cup, \cap, S, \emptyset$⟩ have the same signature because they possess two binary operations ($+$ and $\cdot$) and ($\cup$ and $\cap$) and two constants (1 and 0) and (S and $\emptyset$), respectively.

Example 8.3   The algebras ⟨$\mathbb{Z}, +, 0$⟩ and ⟨$\mathbb{Z}, +$⟩ are not of the same species, because the number of constants is not the same. The second algebra does not possess any constants.

Algebras that have the same signature need not be related at all. In order to be able to distinguish different classes of algebras that "behave" in the same way, we need certain rules that are valid for the elements of the carrier. Such "rules" are called *axioms* and are written as equations of elements of the carrier.

Definition 8.3 (Variety). A set of axioms for the elements of the carrier, together with a signature, specifies a class of algebras called *variety*.

Algebras that belong to the same variety behave in exactly the same way. Although the carrier set, operations and constants may be different, all algebras of the same variety obey the same axioms. In the following sections, we will discuss some of the more important varieties of algebras.

Example 8.4 Consider the variety of algebras with the same signature $\langle \mathbb{R}, +, 0 \rangle$ and the following axioms hold: (i) $x + y = y + x$, (ii) $(x + y) + z = x + (y + z)$, and (iii) $x + 0 = 0 + x = x$. Then $\langle \mathbb{Z}, +, 0 \rangle$, $\langle \wp(S), \cup, \emptyset \rangle$, $\langle \wp(S), \cap, S \rangle$, and $\langle \mathbb{Z}, \circ, 1 \rangle$ are all members of this variety, where the binary operations are denoted as $+$, $\cup$, $\cap$, and $\circ$, and the constants are $0, \emptyset, S$, and 1, respectively. Any theorem proven for this variety will hold for all algebras that belong to this variety.

For the remainder of this chapter, whenever we deal with an arbitrary algebra A, we will use the following notation $A = \langle S, \circ, \Delta, k \rangle$, where S is the carrier, $\circ$ denotes a binary operation, Δ denotes a unary operation, and k is a constant.

8.1.2 Identity and Zero Elements

Constants possess special properties relative to one or more operations in an algebra. The following definition describes the most important properties of constants for binary operations.

Definition 8.4 (Identity and Zero Element). Let $\circ$ be a binary operation on S. An element $1 \in S$ is an *identity* (or *unit*) for the operation $\circ$ if $\forall x \in S$, $1 \circ x = x \circ 1 = x$. An element $0 \in S$ is a *zero* for the operation $\circ$ if $\forall x \in S$, $0 \circ x = x \circ 0 = 0$. If no confusion can result, we may not specify the operation and speak of an *identity* (or *identity element*) and a *zero* (or *zero element*).

Example 8.5 The algebra $\langle \mathbb{Z}, \cdot, 1, 0 \rangle$ with the multiplication as operation has an identity 1 and a zero 0.

Example 8.6 The algebra $\langle \mathbb{Z}, +, 0 \rangle$ has an identity 0 but no zero element.

If identities exist, we can define the inverse with respect to an operation.

Let $\circ$ be a binary operation on S, and 1 an identity for this operation. If $x \circ y = y \circ x = 1$ for every y in S, then x is called (two-sided) *inverse* of y with respect to the operation $\circ$.

Example 8.7 The algebra $\langle \mathbb{Z}, +, 0 \rangle$ has an identity 0 and every element $x \in \mathbb{Z}$ has an inverse with respect to the addition. The inverse of x is written as $-x$ and $x + (-x) = 0$.

Example 8.8 The algebra $\langle \mathbb{R}, \cdot, 1 \rangle$ has an identity 1 and all elements x of the real numbers except 0 have an inverse $x^{-1} = \dfrac{1}{x}$ such that $x \cdot \dfrac{1}{x} = 1$.

8.2   Varieties of Algebras

Algebras play an important role in many applications of computer science such as formal languages and automata theory as well as coding theory and switching theory. In spatial analysis map algebra, i.e., operations on data sets (usually raster), are very common. In this section, we will discuss a few algebras of importance.

8.2.1   Group

Many algebraic structures are the basis of arithmetic, as we usually know it for numbers (integers, rational and real numbers). One basic structure is a group that formalizes the arithmetic of one binary operation (usually addition or multiplication for number sets).

Definition 8.5 (Group). A *group* is an algebra with the signature $\langle S, \circ, ^-, 1 \rangle$ with one binary operation $\circ$ and one unary operation $^-$, where $^-$ is the inverse with respect to $\circ$, and the following axioms hold:
(1) $a \circ (b \circ c) = (a \circ b) \circ c$.
(2) $a \circ 1 = 1 \circ a = a$.
(3) $a \circ \bar{a} = 1$.

If the operation $\circ$ is also commutative, then we call the group a *commutative group* (or *Abelian group*).

Example 8.9 The algebra $\langle \mathbb{Z}, +, -, 0 \rangle$ is a commutative group, where $\mathbb{Z}$ are the integers, $+$ denotes the usual addition, $-$ denotes the inverse (negative number) with regard to the addition, and 0 is the identity for the addition.

The axioms can be easily verified as follows:
(1) $a + (b + c) = (a + b) + c$.
(2) $a + 0 = 0 + a = a$.
(3) $a + (-a) = 0$.
(4) $a + b = b + a$.

Example 8.10 The algebra $\langle \mathbb{R} \setminus \{0\}, \cdot, ^{-1}, 1 \rangle$ is a commutative group, where $\mathbb{R}$ are the real numbers, $\cdot$ is the usual multiplication, $^{-1}$ is the inverse with regard to the multiplication, and 1 is the identity for the multiplication. The axioms are verified as follows:
(1) $a \cdot (b \cdot c) = (a \cdot b) \cdot c$.
(2) $a \cdot 1 = 1 \cdot a = a$.
(3) $a \cdot a^{-1} = 1$.
(4) $a \cdot b = b \cdot a$.

Example 8.11 The natural numbers $\mathbb{N}$ with addition and multiplication are not a group because there is no inverse with regard to addition and multiplication.

8.2.2 Field

Fields are very general algebras that formally describe the interrelation of two binary operations on a carrier set. Simply speaking, a field guarantees all arithmetic operations (as we know them for instance from the usual number sets) without restrictions (except division by zero).

Definition 8.6 (Field). A *field* is an algebra with the signature $\langle S, +, \circ,$ $^-, ^{-1}, 0, 1 \rangle$, where $^-$ and $^{-1}$ are the inverse operations for $+$ and $\circ$, respectively; and the following axioms hold:
(1) $\langle S, +, ^-, 0 \rangle$ is a commutative group.
(2) $a \circ (b \circ c) = (a \circ b) \circ c$.
(3) $a \circ (b + c) = a \circ b + a \circ c$.
(4) $(a + b) \circ c = a \circ c + b \circ c$.
(5) $\langle S \setminus \{0\}, \circ, ^{-1}, 1 \rangle$ is a commutative group.

Example 8.12   The real numbers $\langle \mathbb{R}, +, \cdot, ^-, ^{-1}, 0, 1 \rangle$ are a field with addition and multiplication as binary operations, and the inverse unary operations for addition and multiplication. The numbers 0 and 1 function as identity elements for $+$ and $\cdot$, respectively. The axioms are verified as follows:
(1) Compare Example 8.9 for integers.
(2) See Example 8.10.
(3) $a \cdot (b + c) = a \cdot b + a \cdot c$ (distributive law).
(4) $(a + b) \cdot c = a \cdot c + b \cdot c$ (distributive law).

8.2.3   Boolean Algebra

Definition 8.7 (Boolean Algebra).   A *Boolean algebra* has a signature $\langle S,$ $+, \circ, ^-, 0, 1 \rangle$, where $+$ and $\circ$ are binary operations, and $^-$ is a unary operation (the complementation), with the axioms:
(1) $a + b = b + a$.
(2) $a \circ b = b \circ a$.
(3) $(a + b) + c = a + (b + c)$.
(4) $(a \circ b) \circ c = a \circ (b \circ c)$.
(5) $a \circ (b + c) = a \circ b + a \circ c$.
(6) $a + (b \circ c) = (a + b) \circ (a + c)$.
(7) $a + 0 = a$.
(8) $a \circ 1 = a$.
(9) $a + \bar{a} = 1$.
(10) $a \circ \bar{a} = 0$.

Example 8.13   The power set $\wp(A)$ of a given set A with the usual set operations of union, intersection and complement relative to A is a Boolean algebra $\langle \wp(A), \cup, \cap, ^-, \emptyset, A \rangle$. Let X, Y and Z be arbitrary subsets of A (i.e., elements of the power set of A) then the axioms can easily be verified as:
(1) $X \cup Y = Y \cup X$.
(2) $X \cap Y = Y \cap X$.
(3) $(X \cup Y) \cup Z = X \cup (Y \cup Z)$.
(4) $(X \cap Y) \cap Z = X \cap (Y \cap Z)$.
(5) $X \cap (Y \cup Z) = (X \cap Y) \cup (X \cap Z)$.
(6) $X \cup (Y \cap Z) = (X \cup Y) \cap (X \cup Z)$.

(7) $X \cup \emptyset = X$.
(8) $X \cap A = X$.
(9) $X \cup \overline{X} = A$.
(10) $X \cap \overline{X} = \emptyset$.

8.2.4 Vector Space

Some algebraic structures are defined on more than one set. Vector spaces are one example.

Definition 8.8 (Vector Space). Let $\langle V, +, ^-, 0 \rangle$ be a commutative group and $\langle S, +, \circ, ^-, ^{-1}, 0, 1 \rangle$ a field. V is called a *vector space* over S, if for all $\boldsymbol{a}, \boldsymbol{b} \in V$ and $\alpha, \beta \in S$:
(1) $\alpha \cdot (\boldsymbol{a} + \boldsymbol{b}) = \alpha \cdot \boldsymbol{a} + \alpha \cdot \boldsymbol{b}$.
(2) $(\alpha + \beta) \cdot \boldsymbol{a} = \alpha \cdot \boldsymbol{a} + \beta \cdot \boldsymbol{a}$.
(3) $(\alpha \cdot \beta) \cdot \boldsymbol{a} = \alpha \cdot (\beta \cdot \boldsymbol{a})$.
(4) $1 \cdot \boldsymbol{a} = \boldsymbol{a}$.

The elements of V are called *vectors*; the elements of S are called *scalars*.

Example 8.14 The set of all vectors with $+$ as the vector addition is a vector space over the real numbers where $\cdot$ is the multiplication of a vector with a scalar.

Example 8.15 The set of all matrices with the matrix addition is a vector space over the real numbers with $\cdot$ being the multiplication of a matrix with a scalar.

Vector spaces play an important role in the mathematical discipline of linear algebra, a sub-discipline of algebra.

8.3 Homomorphism

Sometimes, we need to compare algebras to find out whether they are similar. If two algebras are similar they show the same "behavior" in terms of their operations and they have corresponding constants. Often, we know an algebra very well; i.e., we have established theorems for this algebra. If we can show that a different algebra is related to the given algebra (usually we want to show that they are essentially the same in terms of their behavior), then the same theorems (in a related way) also hold for the new algebra.

A formal way of investigating related algebras is to establish a structure-preserving mapping from the (given) algebra to the new algebra. Such a mapping is called a homomorphism.

Definition 8.9 (Homomorphism and Isomorphism). Let $A = \langle S, \circ, \Delta, k \rangle$ and $A' = \langle S', \circ', \Delta', k' \rangle$ be algebras with the same signature, and let h be a function such that

(1) $h : S \to S'$.
(2) $h(a \circ b) = h(a) \circ' h(b)$.
(3) $h(\Delta(a)) = \Delta'(h(a))$.
(4) $h(k) = k'$.

Then h is called a *homomorphism* from A to A'. If the function h is bijective then we call it an *isomorphism* from A to A', and A' is an isomorphic image of A under the map h.

In the definition above $\circ$ and $\circ'$ represent binary operations, Δ and Δ' represent unary operations, and k and k' are constants.

Two algebras that are isomorphic are essentially the same algebra with different names. A homomorphic image of an algebra is a "smaller" or "generalized" version of the given algebra.

Example 8.16 Let S be a nonempty set and $A = \langle \wp(S), \cup, \cap, ^-, \emptyset, S \rangle$ and $B = \langle \{0, 1\}, +, \cdot, ^-, 0, 1 \rangle$ be two Boolean algebras. For any $a \in S$ and $T \in \wp(S)$ the function $h : \wp(S) \to \{0, 1\}$ defined as

$$h(T) = \begin{cases} 0 & \text{if } a \notin T \\ 1 & \text{if } a \in T \end{cases}$$

is a homomorphism from A to B. Note that $h(\emptyset) = 0$ and $h(S) = 1$.

Example 8.17 Let $\mathbb{R}^+$ be the set of all positive real numbers. Then $\langle \mathbb{R}^+, \cdot, 1 \rangle$ is isomorphic to $\langle \mathbb{R}, +, 0 \rangle$ and the function $h : \mathbb{R}^+ \to \mathbb{R}$ with $h(x) = \log_a x$ is an isomorphism. The function h is surjective, because for $x > 0$ the equation $\log_a x = y$ always has a solution $x = a^y$. The logarithmic function is monotone increasing; therefore h is injective. Furthermore, $h(m \cdot n) = \log_a(m \cdot n) = \log_a m + \log_a n = h(m) + h(n)$ and $h(1) = \log_a 1 = 0$. The given isomorphism is the mathematical basis for the slide rule that replaces multiplication of numbers by addition of their logarithms. A slide rule is a calculation device that was frequently used before we had pocket calculators.

8.4   Applications in GIS

Perhaps the most prominent application of algebras in GIS is the *map algebra*. The carrier set of the map algebra is the set of "maps", i.e., data sets that are often referred to by coverage, shapefile, feature class, grid or layer.

We know many operations for manipulating maps. They range from simple arithmetic operations of addition, subtraction, multiplication or division to more complex operations of calculating slope, aspect or hill shade.

An example of a constant of the map algebra would be the zero grid, where every grid cell carries the value zero. Figure 8.1 shows the user interface of the ArcGIS Pro Spatial Analyst raster calculator. Here, we see various arithmetic and logical operators as well as functions that can be applied to map layers.

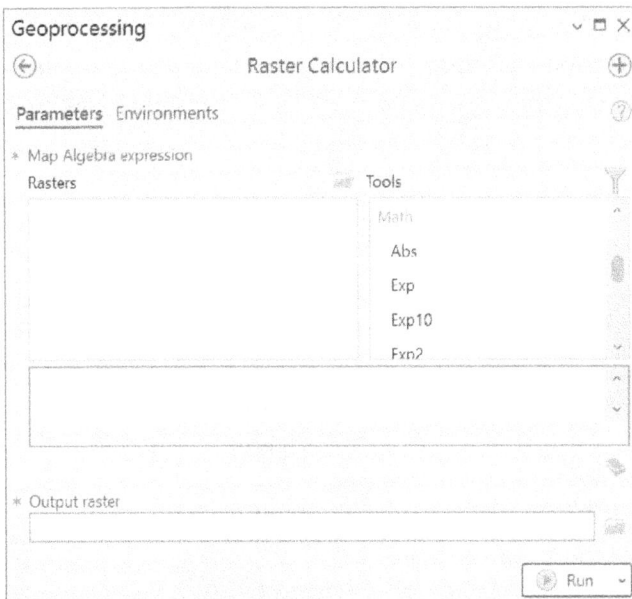

Figure 8.1   ESRI ArcGIS Pro Raster Calculator interface

The concept of structure-preserving mapping finds an application in spatial modeling, where we map a subset of the real world to a representation in a spatial feature model.

8.5 Exercises

Exercise 8.1 Given two algebras: $\langle \mathbb{Z}, +, -, 0 \rangle$ with the integers as carrier set, two operations, addition $+$ and negative number $-$, and the constant 0; $\langle \boldsymbol{E}, +, -, 0 \rangle$ with the even numbers as carrier set, the constant 0 and the operations $+$ and $-$ defined in the usual way (addition and negative number, respectively). Show that both algebras are Abelian groups and that the function $f : \mathbb{Z} \to \boldsymbol{E}$, defined as $f(x) = 2x$, is an isomorphism.

Exercise 8.2 Let $\{T, F\}$ be a set where T stands for "true" and F stands for "false". Show that $\langle \{T, F\}, \vee, \wedge, \neg, F, T \rangle$ is a Boolean algebra with the binary operations "and" $\wedge$, "or" $\vee$, and the unary operation "not" $\neg$ being logical operators. The constants F and T are propositions that are always false F and always true T, respectively.

Exercise 8.3 Given two algebras: $\langle \mathbb{Z}, +, -, 0 \rangle$ with the integers as carrier set, two operations, addition $+$ and negative number $-$, and the constant 0; $\langle B, +, -, 0 \rangle$ with the carrier set $B = \{0, 1\}$, the constant 0 and two operations defined as

$$
\begin{array}{c|cc}
+ & 0 & 1 \\
\hline
0 & 0 & 1 \\
1 & 1 & 0
\end{array}
$$

and $(-x) = x$. Show that the function $f : \mathbb{Z} \to B$, defined as

$$
f(x) = \begin{cases} 0 & x \text{ is even} \\ 1 & x \text{ is odd} \end{cases}
$$

is a homomorphism. Why is f not an isomorphism?

Chapter 9

Topology

Topology is a central concept in every GIS. It deals with the structural representation of spatial features and their properties that remain invariant under certain transformations. In this chapter, we introduce the mathematical concept of a topological space based on the topology that is induced on the real plane by a distance function.

We also show how simple structures can be used to build complex objects in GIS databases, and how to check the consistency of a two-dimensional topologic representation of spatial features.

9.1 Topological Spaces

In this section, we will deal with topological spaces, i.e., a set and a collection of subsets of this set that satisfy certain conditions. There are two equivalent approaches to define a topological space. The first one starts with the concept of a neighborhood of a point and defines a topological space as a system of neighborhoods that fulfill certain conditions. The concept of open sets follows from the definition of a neighborhood. The second approach starts from a family of subsets of a given set (which are called open sets) and defines a topology through properties of these open sets. The concept of a neighborhood follows from the definition of a topological space.

9.1.1 Metric Spaces and Neighborhoods

The first approach is more intuitive than the second one that is usually used in general topology (or point-set topology). For our purpose, we chose the intuitive approach on the Euclidean plane with a system of neighborhoods. In order to define a neighborhood, we need the concept of a distance. Generally, this can be achieved with a metric space.

Definition 9.1 (Metric Space). Let X be a nonempty set and d a function $X \times X \rightarrow \mathbb{R}_0^+$ (the positive real numbers including zero) such that for every $x, y, z \in X$:
(1) $d(x, y) = 0$ if and only if $x = y$.
(2) $d(x, y) = d(y, x)$.
(3) $d(x, y) + d(y, z) \geq d(x, z)$ (triangle inequality).

We call the pair (X, d) a *metric space* and d a *distance function* (or *metric*) on X.

Example 9.1 Let us consider the real plane $\mathbb{R}^2$ equipped with the *Euclidean distance* $d_E(p, q) = \sqrt{(a_1 - b_1)^2 + (a_2 - b_2)^2}$ between two points $p = (a_1, a_2)$ and $q = (b_1, b_2)$. We call this space the two-dimensional *Euclidean space*. $(\mathbb{R}^2, d_E)$ is a metric space. The Euclidean distance is the shortest distance between two points. This is the usual space of plane geometry. We can easily extend this space to three dimensions.

Example 9.2 The real numbers $\mathbb{R}$ with the distance function $d(x, y) = |x - y|$ are a metric space.

In every metric space, we can define a neighborhood for points of this space.

Definition 9.2 (ε-Neighborhood). In a metric space (X, d), for each $x \in X$ and each $\varepsilon > 0$, we define an (open) ε-*neighborhood* of x as the set $N(x, \varepsilon) = \{y | y \in X \wedge d(x, y) < \varepsilon\}$. When no confusion is possible we call $N(x, \varepsilon)$ *neighborhood* and write $N(x)$.

The set $N_d(x) = \{N(x, \varepsilon) | x \in X \wedge \varepsilon > 0\}$ is called the *neighborhood system* of x induced by the metric d. In short, we will write $N(x)$.

In the Euclidean plane $\mathbb{R}^2$ an open disk with radius ε around a point p is an ε-neighborhood (Figure 9.1).

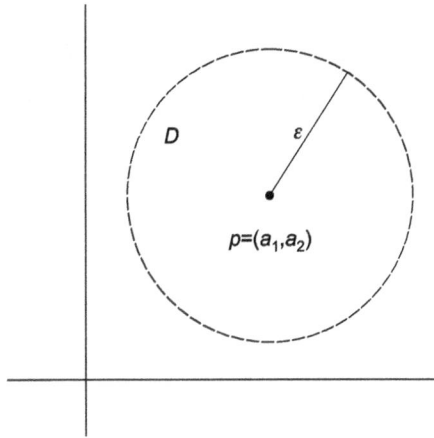

Figure 9.1   Open disk in $\mathbb{R}^2$

9.1.2   Topology and Open Sets

When we require a neighborhood system to satisfy certain conditions, we arrive at the definition of a topological space.

Definition 9.3 (Topological Space). Let X be a set and for every $x \in X$ there exists a neighborhood system $\mathcal{N}(x) \subseteq \wp(X)$ that satisfies the following conditions (neighborhood axioms):

(N1) The point x lies in each of its neighborhoods.

(N2) The intersection of two neighborhoods of x is itself a neighborhood.

(N3) Every superset U of a neighborhood N of x is a neighborhood of x. X is a neighborhood of x.

(N4) Every neighborhood N of x contains a neighborhood V of x such that N is a neighborhood of every point of V.

We then call the neighborhood system $\mathcal{N}(x)$ a *topology* on X, and the set with its neighborhood system $(X, \mathcal{N}(x))$ a *topological space*. Sometimes we denote a topological space simply by X.

Figure 9.2 illustrates the four neighborhood axioms.

With the help of the concept of a neighborhood, we can now define open sets.

Definition 9.4 (Open Set). Let X be a topological space. A subset O of X is an *open set* if it is a neighborhood for each of its points.

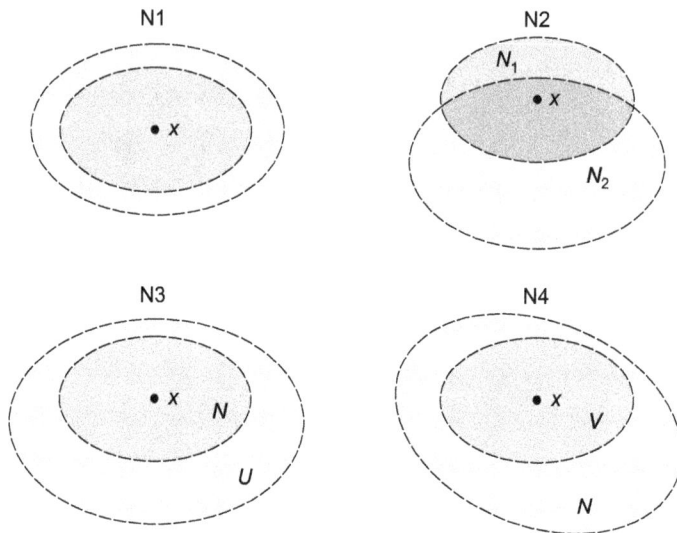

Figure 9.2   Neighborhood axioms

Example 9.3   The open interval (a, b) in the real numbers and the open disks in $\mathbb{R}^2$ are open sets.

Definition 9.5 (Closed Set). Let X be a topological space. A set C is *closed* if its complement $X - C$ is open.

Example 9.4   The closed interval $[a, b]$ in the real numbers is a closed set, because its complement $(-\infty, a) \cup (b, \infty)$ is the union of two open sets, which again is an open set.

Example 9.5   The half-open interval $(a, b]$ in the real numbers is neither open nor closed.

The previous example shows that "closed" does not mean "not open". Sets can be neither open nor closed, or they can be both open and closed (sometimes called *clopen* sets). The following statements can be proven to be true for open sets.

(O1) The empty set $\emptyset$ and the set X are open.

(O2) The intersection of any finite number of open sets is open.

(O3) The union of any number of open sets is open.

(O4) A subset U of X is a neighborhood of $x \in X$ if and only if there exists an open set O with $x \in O \subseteq U$.

The intersection of an arbitrary number of open sets does not need to be open. Take example for the intersection of an infinite collection of open intervals

$$\left(-\frac{1}{n}, +\frac{1}{n}\right), \quad n = 1, 2, 3, \ldots$$

Obviously, the intersection is the set $\{0\}$ which is not an open set.

It can be proven that the intersection of an arbitrary number of closed sets, and the union of a finite number of closed sets, are closed.

Our approach to the definition of a topological space is based on the concept of the ε-neighborhood defined in a metric space. For this definition we need the distance, a concept which is too special for general topological spaces. Statement (O4) above gives us a way to define neighborhoods without the notion of distance. Here, we also see that a neighborhood does not need to be an open set; it can also be a closed set. As an example, consider a point p of the Euclidean plane $\mathbb{R}^2$. Every closed disk around p is a neighborhood of p, because it contains the open disk around p, which is an open set.

9.1.3 Continuous Functions and Homeomorphisms

We can define mappings between topological spaces. A function that maps the neighborhood of a point to the neighborhood of the image of this point is called a continuous function. We can define it formally as follows.

Definition 9.6 (Continuous Function). Let $f : X \to Y$ be a function from the topological space X to the topological space Y. We call f *continuous* at point $x_0 \in X$ if for every neighborhood V of $f(x_0)$ there is a neighborhood U of x_0 such that the image of U, i.e., $f(U)$, is a subset of V. If f is continuous at every point of X, we call it a *continuous function*.

Figure 9.3 illustrates both concepts of continuity at a point and continuous function.

The definition above is valid for any two topological spaces. For the real numbers $\mathbb{R}$, the definition of continuity usually reduces to the following statement:

A function $f : \mathbb{R} \to \mathbb{R}$ is continuous at point x_0 if for every $\varepsilon > 0$ there exists a $\delta > 0$ such that $|x - x_0| < \delta$ implies $|f(x) - f(x_0)| < \varepsilon$. A function is continuous if it is continuous at every point. Continuity of a

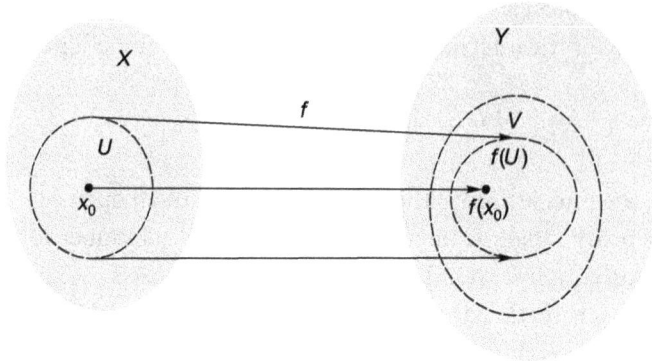

Figure 9.3 Continuous function

function essentially means that the graph of the function has no "jumps" or "gaps".

Like in other mathematical structures, also for topological spaces we know structure-preserving mappings. They map one topological space to another topological space thereby preserving the topology.

Definition 9.7 (Homeomorphism). Let $h : X \to Y$ be a function from the topological space X to the topological space Y. If this function is continuous, bijective, and possesses a continuous inverse, we call it a *homoeomorphism* (or *topological mapping*).

If two spaces are homeomorphic, they are essentially the same and expose the same topological behavior.

Example 9.6 Let $X = (-1, 1)$ be an open interval in $\mathbb{R}$ and $f : X \to \mathbb{R}$ a function defined as $f(x) = \tan \dfrac{\pi}{2} x$. This function is bijective, continuous and has a continuous inverse. Figure 9.4 shows the graph of the function. It is a homeomorphism. This means that the open interval $(-1, 1)$ and the real numbers are homeomorphic.

Example 9.7 The open disk $D_1 = \{(r, \theta) | r < 1\}$ given with its polar coordinates (r, θ) and radius 1 is homeomorphic to the open disk $D_2 = \{(r, \theta) | r < 2\}$ with radius 2 through the function $f(r, \theta) = (2r, \theta)$.

A property of a topological space that is preserved by a homeomorphism is called a *topological property* or a *topological invariant*. Mathematical topology

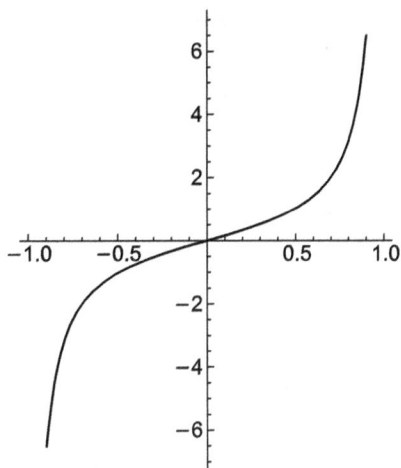

Figure 9.4   Example of a homeomorphic function

is mainly focused on properties of topological spaces that remain invariant under topological mappings.

The two previous examples show that length and area are not topological invariants, because the length of the interval $(-1, 1)$ is different from the "length" of the real line, and in the second example both disks are homeomorphic but their areas are not the same.

9.1.4   Alternate Definition of A Topological Space

As mentioned earlier, a different, yet equivalent, definition of a topological space starts with the idea of an open set, defines a topology as properties of a collection of open sets and derives the definition of a neighborhood from the open sets. We take the properties O1 to O3 of open sets as axioms and define a topological space as follows.

Definition 9.8 (Topological Space). Let X be a set and O be a collection of subsets of X, i.e., $O \subseteq \wp(X)$. We call O a *topology* on X when the subsets satisfy the following three conditions:
(O1) $\emptyset \in O, X \in O$.
(O2) $A, B \in O \Rightarrow A \cap B \in O$.
(O3) $A_i \in O \Rightarrow \bigcup_{\forall i \in I} A_i \in O$.

We call the O_i *open sets*, (X, O) a *topological space* and the elements $x \in X$ the *points* of the topological space.

The three conditions for a topology require that the empty set and the set itself must always be a member of the topology. Further, the intersection of a finite number of open sets is always an open set, and the union of an arbitrary number of open sets is an open set.

With the help of the open sets used in the definition of a topology we can now define a neighborhood.

Definition 9.9 (Neighborhood). N is a *neighborhood* of the point x if $N \subseteq X$ and there exists an open set $A \in O$ such that $x \in A \subseteq N$.

With this definition, we can prove the statements N1 to N4 of Definition 9.3 about neighborhoods to be true.

Example 9.8 Two extreme topologies can be found on any set X. The first one consists only of two elements $\{X, \emptyset\}$, the second one consists of all subsets of X, i.e., the power set $\wp(X)$. We can easily verify that the three conditions are satisfied for both topologies. The first topology is called *indiscrete topology*. It is the coarsest of all topologies, because it consists of only two elements. The second one is called *discrete topology*, which is the finest of all topologies.

Example 9.9 Consider the real line $\mathbb{R}^1$. We call a subset A of $\mathbb{R}^1$ an open set if it is empty or with each of its points $x \in A$ contains an open interval S_x that completely lies within A. All open intervals (a, b) on the real line $\mathbb{R}^1$ are open sets. The real line itself is an open set. Again, we can show that these open sets are a topology on $\mathbb{R}^1$. We call it the *natural topology*. This can be extended to the $\mathbb{R}^n$ with open disks, balls, etc.

Both approaches to the definition of a topological space as mentioned above are equally valid and lead to the same results. Figure 9.5 summarizes both approaches: the intuitive approach based on the concept of a neighborhood, and the set theoretic abstract approach based on the concept of open sets.

The first one defines a topological space through properties of neighborhoods. Open sets are then defined through neighborhoods, and the properties O1 to O4 follow as theorems.

The second approach defines a topological space through properties of open sets. It then defines neighborhoods through property O4 of open sets,

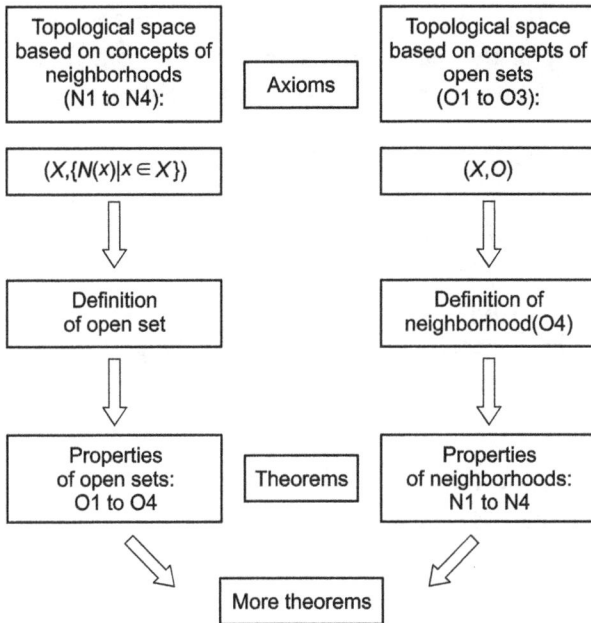

Figure 9.5 Equivalent approaches to the definition of a topological space, open sets and neighborhoods and related theorems

and derives N1 to N4 as theorems about neighborhoods. More theorems about topological spaces follow from there.

9.2 Base, Interior, Closure, Boundary, and Exterior

From the definition of a topological space we know that the union of open sets is an open set. If in a topological space every open set can be generated as the union of some open sets, we call these sets a base of the topology.

Definition 9.10 (Base). Let X be a topological space and $\mathcal{B}$ be a collection of open sets such that every open set of the topology is the union of members of $\mathcal{B}$. Then $\mathcal{B}$ is called a *base* for the topology and the elements of $\mathcal{B}$ are called *basic open sets*.

An equivalent definition for a base requires that for every point $x \in X$ that belongs to an open set O there is always an element $B \in \mathcal{B}$ such that $x \in B \subset O$.

Example 9.10   The open disks in the Euclidean plane $\mathbb{R}^2$ are a base for the natural topology of the plane. This follows easily from the definition of a neighborhood. The number of the basic open sets is uncountably infinite.

Example 9.11   The open disks in the Euclidean plane $\mathbb{R}^2$ with radii and center coordinates being rational numbers are a base for the natural topology of the plane. Note that the number of the basic open sets is countably infinite.

Whereas a base for a topology is a global characteristic of a topological space, we can also define a local base at a point of a topological space. This is a local characteristic of a topological space determined only by the neighborhood of the point.

Definition 9.11 (Local Base).   A collection $\mathcal{B}$ of neighborhoods of a point x of a topological space X is a *local base* at x if every neighborhood of x contains some members of $\mathcal{B}$.

Example 9.12   Consider the natural topology in the Euclidean plane $\mathbb{R}^2$ and a point x. The system of open disks $\mathcal{B}_x$ with center x is a local base at x. This is true because for every open set O that contains x there is an open disk with center at x that is contained in O.

Example 9.13   Let x be a point of a metric space. The countably infinite set of ε-neighborhoods of x defined as $\left\{ N(x,1), N\left(x, \frac{1}{2} \right), N\left(x, \frac{1}{3} \right), ... \right\}$ is a local base at x.

The relationship between a base of a topology and a local base at a point can be expressed in the following statement:

Let $\mathcal{B}$ be the base for a topology and $x \in X$ be a point of the topological space. Then the members of $\mathcal{B}$ that contain x form a local base at x.

For further investigations, we need the concepts of interior, closure, boundary and exterior of a set.

Definition 9.12 (Interior, Closure, Boundary, and Exterior).   Given a subset A of a topological space X we define the interior, closure and boundary as follows:

- The union of all open sets contained in A is called the *interior* of A (written as A°).

- The smallest closed set containing A is called the *closure* of A (written as $\overline{A}$), in other words it is the intersection of all closed sets containing A.[1]
- The *boundary* of set A is the intersection of the closure of A with the closure of its complement $X - A$. The boundary[2] is written as ∂A.
- The *exterior* of a set A (written as A^-) is the interior of the complement of A, i.e., $A^- = (X - A)^\circ$.

An open set is its own interior. A closed set is equal to its closure. Table 9.1 shows some properties of interior, closure, and boundary.

Table 9.1   Properties of interior, closure, and boundary of a set

Interior	Closure	Boundary
$A^\circ \subseteq A, (A^\circ)^\circ = A^\circ$	$A \subseteq \overline{A}, \overline{\overline{A}} = \overline{A}$	$\partial A = \overline{A} - A^\circ$
$A \subseteq B \Rightarrow A^\circ \subseteq B^\circ$	$A \subseteq B \Rightarrow \overline{A} \subseteq \overline{B}$	$\partial A = \overline{A} \cap (X - A^\circ)$
$(A \cap B)^\circ = A^\circ \cap B^\circ$	$\overline{A \cup B} = \overline{A} \cup \overline{B}$	$\partial A = \overline{A} \cap \overline{X - A}$
$\left(\bigcup_{i \in I} A_i\right)^\circ \supseteq \bigcup_{i \in I} A_i^\circ$	$\overline{\bigcup_{i \in I} A_i} \supseteq \bigcup_{i \in I} \overline{A_i}$	$\partial A = A - (A^\circ \cup (X - A)^\circ)$
$\left(\bigcap_{i \in I} A_i\right)^\circ \subseteq \bigcap_{i \in I} A_i^\circ$	$\overline{\bigcap_{i \in I} A_i} \subseteq \bigcap_{i \in I} \overline{A_i}$	$\partial A = \partial(X - A)$

Example 9.14   Consider the set $X = \{a, b, c, d, e\}$, the topology O defined on X as $O = \{X, \emptyset, \{a\}, \{c, d\}, \{a, c, d\}, \{b, c, d, e\}\}$ and the subset $A = \{b, c, d\}$ of X. The interior of A is $A^\circ = \{c, d\}$, because the only open sets contained in A are $\{c, d\}$ and $\emptyset$ whose union is $\{c, d\}$. The closure of A is $\overline{A} = \{b, c, d, e\}$, because among the closed sets[3] of X, i.e., $\emptyset, X, \{b, c, d, e\}$, $\{a, b, e\}, \{b, e\}$ and $\{a\}$, the smallest one that contains A is $\{b, c, d, e\}$. The boundary of A is the difference of the closure with the interior, i.e., $\partial A = \overline{A} - A^\circ = \{b, c, d, e\} - \{c, d\} = \{b, e\}$. The exterior of A is the interior of the complement of A, i.e., $\{a, e\}^\circ$, which results to $\{a\}$.

Example 9.15   Let us consider an open subset A of the Euclidean plane $\mathbb{R}^2$. Figure 9.6 illustrates the interior, boundary, closure and exterior of the set.

---

[1]Note that we use for closure the same symbol as for the set complement. They are, however, not related to each other.

[2]The boundary of a set is often denoted as *frontier* of a set.

[3]The closed sets are the complements of the open sets.

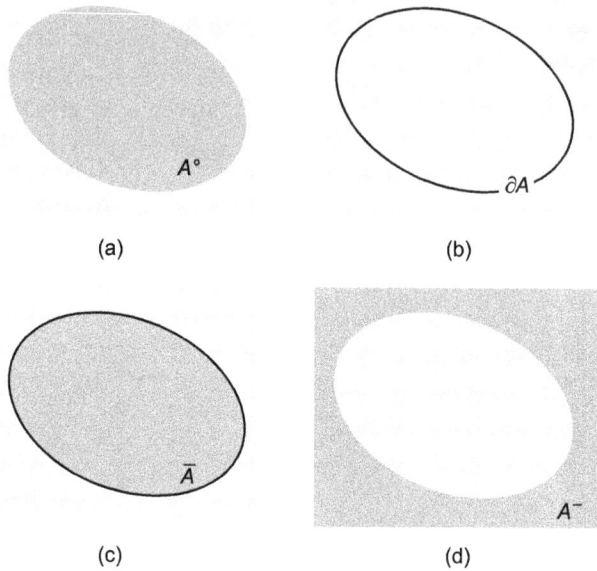

Figure 9.6   (a) Interior, (b) boundary, (c) closure and (d) exterior of an open set

9.3   Classification of Topological Spaces

There are several ways to classify topological spaces. Usually, this is done according to the degree to which their points are separated, regarding their compactness, their overall size, and their connectedness. Let us look at each of them in some detail.

9.3.1   Separation Axioms

In topology, we know many different ways to distinguish disjoint sets and distinct points. We first present axioms that separate two distinct points.

Definition 9.13 (T_0 Space). We call a topological space T_0 (or a T_0 space) if for two distinct points at least one has a neighborhood that does not contain the other point.

Definition 9.14 (T_1 Space). We call a topological space T_1 (or a T_1 space) if two distinct points have neighborhoods that do not contain the other point.

Definition 9.15 (Hausdorff Space). A topological space X is called a Hausdorff space or T_2 space if two distinct points $a, b \in X$ possess disjoint open neighborhoods. In other words, there exist two open sets A and B with $a \in A$ and $b \in B$ and $A \cap B = \emptyset$.

Every metric space with the metric topology is T_2. Hausdorff spaces are always T_1, and every T_1-space is always T_0.

Figure 9.7 illustrates separation axioms T_0, T_1, and T_2.

We now look at axioms that separate sets. First, we define an axiom that separates closed sets from the points that lie outside that set.

Definition 9.16 (Regular Space). If a topological space is T_1 and for every closed set C and every point x outside C there exist an open set A that contains C and a disjoint neighborhood N of x, then we call this space a regular space or T_3 *space*.

Every metric space with the metric topology is regular. Every regular space is a Hausdorff space. The converse is not true, because there are Hausdorff spaces that are not regular.

Finally, we introduce a separation axiom that separates closed sets.

Definition 9.17 (Normal Space). If a topological space is T_1 and for any two disjoint closed sets C_1 and C_2 there exist disjoint neighborhoods that contain the closed sets, then we call this space a normal space or T_4 space.

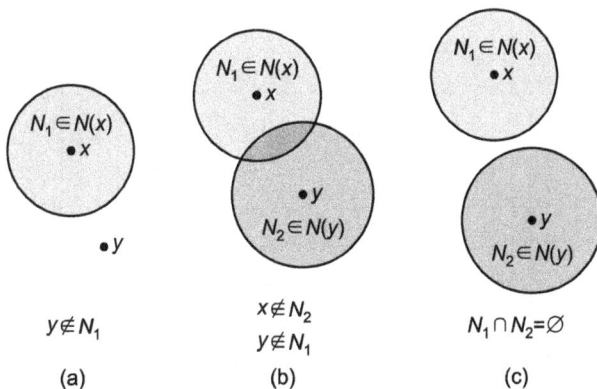

Figure 9.7 Separation axioms T_0, T_1, and T_2. (a) T_0 axiom, (b) T_1 axiom and (c) T_2 axiom

Every metric space with the metric topology is normal, and every normal space is regular. The converse is not true, because there exist regular spaces that are not normal.

Figure 9.8 illustrates the axioms that lead to the definition of regular (T_3 and T_1 axioms) and normal spaces (T_4 and T_1 axioms).

Figure 9.9 shows the relations between the separation characteristics of topological spaces.

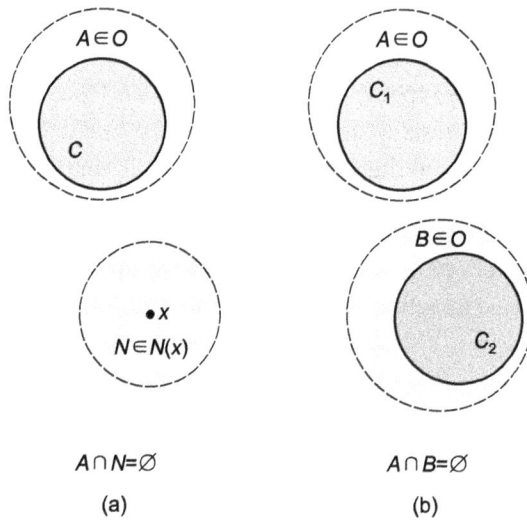

Figure 9.8   Separation axioms T_3 and T_4. (a) T_3 axiom and (b) T_4 axiom

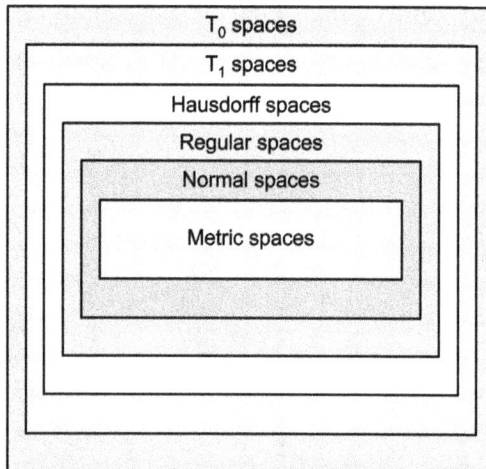

Figure 9.9   Relationship between separation characteristics of topological spaces

Separation characteristics are topological properties of a space, i.e., if a topological space X has a certain separation characteristic and there is a homeomorphism from X to a topological space Y, also Y will have the same characteristic. We see that a metric space is a very special case of a topological space that possesses all separation characteristics.

9.3.2 Compactness

In this section we discuss properties of topological spaces whose conditions are stronger than separation characteristics. These conditions are defined by what are called open covers. Finally, we will see that subsets of a Euclidean space which are both closed and bounded[4] are of special importance.

Definition 9.18 (Open Cover). Let X be a topological space and $\mathcal{F}$ be a family of open subsets of X. If the union of these subsets is the whole space X, we call $\mathcal{F}$ an open cover of X. If $\mathcal{F}'$ is a subfamily of $\mathcal{F}$ with $\cup \mathcal{F}' = X$, then $\mathcal{F}'$ is a subcover of $\mathcal{F}$.

Example 9.16 Consider all open disks in the Euclidean plane $\mathbb{R}^2$ whose radii are 1 and their centers have integer coordinates. The union of these disks covers the whole space. Therefore, it is an open cover of $\mathbb{R}^2$. If we leave out one disk, their union is not the whole space any more. Therefore, this family of open disks has no subcover.

If a topological space has a finite subcover, we call this space compact space.

Definition 9.19 (Compact Space). If every open cover of a topological space X has a finite subcover, we call X a *compact space*.

The following spaces are compact:
- The closed unit interval $[0, 1]$.
- Any finite topological space.
- A closed interval, disk or ball in $\mathbb{R}^1$, $\mathbb{R}^2$, or $\mathbb{R}^3$, respectively.

The following statements can be made about compact spaces. Their proofs can be found in the topological literature:

---

[4] A set is *bounded* when it is contained in some open ball with finite radius.

- A continuous image of a compact space is compact.[5]
- A closed subset of a compact space is compact.
- A subset of the Euclidean n-space is compact if and only if it is closed and bounded.
- A compact Hausdorff space is normal.

The results from the previous section about separation and the last item from the previous list tell us that metric spaces as well as compact Hausdorff spaces are normal.

   If we relax the condition for compactness to have a countable cover instead of a finite cover we arrive at the definition of a Lindelöf space.

Definition 9.20 (Lindelöf Space). If every open cover of a topological space X has a countable subcover, we call X a *Lindelöf space* (or we say that space X is Lindelöf). Compact spaces are always Lindelöf.

Example 9.17   The Euclidean plane $\mathbb{R}^2$ equipped with the natural topology of open disks is a Lindelöf space.

Compactness is a topological property that remains invariant under homeomorphisms.

9.3.3   Size

A further characterization of topological spaces can be done according to their sizes. A measure for the size of a set is its cardinality, or number of elements. We recall from Chapter 5 on set theory that a set is *countable* if it either has a finite number of elements or is countably infinite.

   In order to proceed we need the definition of the term dense.

Definition 9.21 (Dense Set). A subset A of a topological space X is *dense* if its closure is X, i.e., $\overline{A} = X$.

---

[5]This proposition states that compactness as a topological property is even preserved under the weaker condition of a continuous mapping (and not a homeomorphism).

Example 9.18 The rational numbers $\mathbb{Q}$ are a dense subset of the real numbers, because it can be shown that $\overline{\mathbb{Q}} = \mathbb{R}$. The rational numbers are countably infinite.

With both properties of a set to be countable and dense we can impose some limit on the size of a topological space which leads to the definition of a separable topological space.

Definition 9.22 (Separable Space). A topological space is *separable* if it has a countable dense subset.

Example 9.19 The n-dimensional Euclidean space is separable.

Definition 9.23 (First-Countable). A topological space is *first-countable* if every point has a countable local base.

Example 9.20 Every metric space is first-countable. According to Example 9.13, we have identified a countable local base for a metric space. It is therefore first-countable.

Example 9.21 Every discrete topological space is first-countable.

First-countable is a local property of a topological space, which is solely determined by the properties of the neighborhoods of a point. Another property of a topological space is related more to a global characteristic of space.

Definition 9.24 (Second-Countable). A topological space is *second-countable* if it has a countable base for its topology.

Example 9.22 The Euclidean plane $\mathbb{R}^2$ is second-countable. According to Example 9.11 the open disks with rational radii and center coordinates are a countable base for $\mathbb{R}^2$. Therefore, $\mathbb{R}^2$ is second-countable.

For topological spaces we can make the following statements with regard to their size characteristics: If a topological space is second-countable, then it is also first-countable, separable, and Lindelöf.

The properties of a topological space to be separable, first-countable, and second-countable are topological properties and remain invariant under homeomorphisms.

9.3.4 Connectedness

Connectedness of topological spaces deals with the property of such spaces that they cannot be divided into two disjoint nonempty open sets whose union is the entire space.

Definition 9.25 (Connected Space). A space X is *connected* if whenever it is represented as the union of two nonempty subsets $X = A \cup B$ then $\overline{A} \cap B \neq \emptyset$ or $A \cap \overline{B} \neq \emptyset$.

Intuitively speaking, a space is connected if it appears in one piece or it cannot be represented as the union of two disjoint nonempty open subsets. The following conditions on a topological space are equivalent to formulate connectedness:

- X is connected.
- The only subsets of X that are both open and closed are the empty set $\emptyset$ and X.
- X cannot be expressed as the union of two disjoint nonempty open sets.

Example 9.23 The Euclidean space $\mathbb{R}^n$ is connected, because the empty set and $\mathbb{R}^n$ are the only sets that are both open and closed.

Example 9.24 Let $X = \{a, b, c, d, e\}$ be a set and $O = \{X, \emptyset, \{a\}, \{c, d\}, \{a, c, d\}, \{b, c, d, e\}\}$ a topology on X. Then X is not connected, because $\{a\}$ and $\{b, c, d, e\}$ are disjoint open sets and $X = \{a\} \cup \{b, c, d, e\}$ is the union of two disjoint nonempty open subsets.

A somewhat stronger condition can be stated when we consider how two points in a topological space can be connected.

Definition 9.26 (Path-Connected Space). A topological space X is *path-connected* if any two points $x_1, x_2 \in X$ of the space can be connected by a path. A *path* in a topological space X is a continuous function $f : [0, 1] \to X$ such that $f(0) = x_1$ (beginning point) and $f(1) = x_2$ (end point).

In general, every path-connected space is connected. The converse is not true. However, for regions[6] of the Euclidean plane $\mathbb{R}^2$ with the natural

---

[6]An open connected subset of a topological space is called a *region*.

topology we have the following result: Every open connected subset of $\mathbb{R}^2$ is path-connected.

Connectedness is a topological property, i.e., it remains invariant under homeomorphisms. The image of a connected set under a continuous mapping is connected.

9.4 Simplicial Complexes and Cell Complexes

The topological spaces we have treated so far are usually very general and too complex for many investigations related to spatial data. We, therefore, look for simpler spaces that can be used instead. These spaces can then be pieced together to form more complex spaces, yet keeping a recognizable shape and being easy to handle.

One class of these simple spaces is polyhedra. A polyhedron is a topological space that is built of simple building blocks, the simplexes. A generalization of polyhedra leads to cell complexes (or CW complexes) glued together from cells.

9.4.1 Simplexes and Polyhedra

We first need to introduce the concept of a simplex. Simply speaking, a simplex is the simplest geometric figure of a respective geometric dimension in the Euclidean space, i.e., a point in a zero-dimensional space, a straight-line segment in a one-dimensional space, a triangle in a two-dimensional space, and a tetrahedron in a three-dimensional space.

Definition 9.27 (Simplex). Given $k+1$ points $v_0, v_1, \ldots, v_k \in \mathbb{R}^n$ in general position, where $k \leq n$, we call the smallest convex set containing them a *closed k-simplex* (or *simplex of dimension k*), written as $\overline{\sigma}^k$. The points $v_0, \ldots, v_k$ are called the *vertices* of the simplex. A closed simplex can be written as $\overline{\sigma}^k = \lambda_0 v_0 + \lambda_1 v_1 + \cdots + \lambda_k v_k$ where the $\lambda_0, \ldots, \lambda_k \in \mathbb{R}_0^+$ and $\lambda_0 + \lambda_1 + \cdots + \lambda_k = 1$. If we require $\lambda_0, \ldots, \lambda_k \in \mathbb{R}^+$ (positive real numbers excluding zero) we get an *open k-simplex*, written as σ^k.

Figure 9.10 illustrates the definition with closed simplexes of dimensions 0, 1, 2, and 3.

| 0-simplex | 1-simplex | 2-simplex | 3-simplex |
| (point) | (closed line segment) | (triangle) | (solid tetrahedron) |

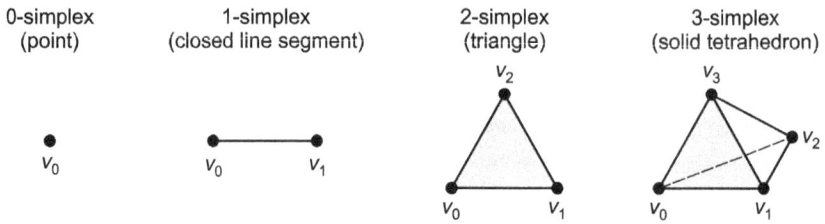

Figure 9.10   Simplexes of dimensions 0, 1, 2, and 3

The convex hull of a nonempty subset of the vertices of a closed simplex $\overline{\sigma}^k$ is called a face of the simplex. If a simplex is of dimension n then a k-face is a simplex of dimension with $k < n$.

In Figure 9.10, the closed 1-simplex has two 0-faces v_0 and v_1. The triangle has three 0-faces, v_0, v_1 and v_2, and three 1-faces, v_0v_1, v_1v_2, and v_2v_0. The tetrahedron has four 0-faces, six 1-faces, and four 2-faces.

A simplex is a topological space with the natural topology derived from its embedding Euclidean space. We can now piece together simplexes in a defined way to a simplicial complex.

Definition 9.28 (Simplicial Complex). A finite collection K of closed simplexes in a Euclidean space $\mathbb{R}^n$ is called a *simplicial complex* if the following two conditions are satisfied:
(1) For every simplex all of its faces must also be in the collection.
(2) If two simplexes intersect then they must do so in a common face.

Figure 9.11 shows two collections of simplexes in the two-dimensional Euclidean space. The one on the left is a simplicial complex. The right one violates the conditions for a simplicial complex. The lower triangle touches the other one on the base. There is, however, no common face. The line segment intersects the upper triangle, but not in a common face.

Note that a simplicial complex is a set of simplexes and as such not a topological space. However, if we consider the union of all simplexes in a simplicial complex as a subset of a Euclidean space, we can apply the subspace topology[7] and make it a topological space. A simplicial complex K, when

---

[7]Given a topological space X and a subset Y of this space, we define a *subspace* by intersecting the open sets of X with Y. This gives us the open sets for a new topology on Y, the *subspace topology*. Y is called a *subspace* of X.

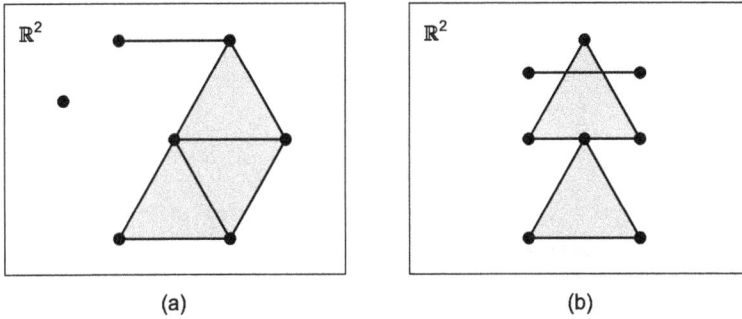

Figure 9.11   Valid simplicial complex (a) and invalid simplicial complex (b)

viewed in this way, is a topological space that we call a polyhedron, written as $|K|$.

Polyhedra possess useful properties. As closed and bounded subsets of a Euclidean space they are compact and metric spaces.

9.4.2   Cells and Cell Complexes

For many topological investigations polyhedra are too special and too complex. On the other hand, general topological spaces are too general. A concept in between that also possesses many useful properties is a cell complex.

Definition 9.29 (Unit Ball, Unit Sphere, Unit Cell, Cell). Let $\mathbb{R}^n$ be the n-dimensional Euclidean space with the natural topology. The subspace $D^n = \{x \in \mathbb{R}^n \| |x| \leq 1\}$ is called the n-dimensional *unit ball*. The $(n-1)$-dimensional subspace $S^{n-1} = \{x \in \mathbb{R}^n \| |x| = 1\}$ is called the $(n-1)$-dimensional *unit sphere*. The subspace $\mathring{D}^n = \{x \in \mathbb{R}^n \| |x| < 1\}$ is called the n-dimensional *unit cell*. A topological space homeomorphic to $\mathring{D}^n$ is called a n-dimensional *cell* (or n-cell).

Example 9.25   In $\mathbb{R}^2$ the unit ball is a closed disk with radius 1, the unit sphere is the circle with radius 1, and the unit cell is the open disk with radius 1.

Example 9.26   Every open n-simplex is a n-cell.

●	0-dimensional unit cell	●	0-cell
----------	1-dimensional unit cell	⌇	1-cell
●	2-dimensional unit cell	⬡	2-cell
●	3-dimensional unit cell	⬡	3-cell

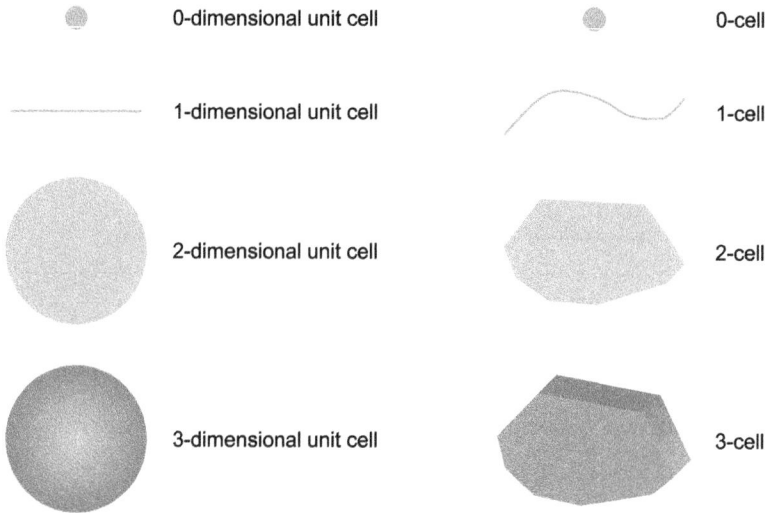

Figure 9.12   Unit balls and cells

Figure 9.12 shows unit balls of dimensions 0, 1, 2, and 3 and corresponding 0-, 1-, 2-, and 3-cells.

We can now generalize the concept of a simplicial complex to a cell complex, which is defined as a collection of cells that are glued together in a certain way.

Definition 9.30 (Cell Decomposition, Skeleton). A *cell decomposition* is a topological space X and a set C of subspaces of X whose elements are cells such that X is the disjoint union of these cells, i.e., $X = \cup_{c \in C} c$. The n-dimensional *skeleton* of X is the subspace $X^n = \cup\{c \in C | dim(c) \leq n\}$. We have then a sequence of subspaces $\emptyset = X^{-1} \subset X^0 \subset X^1 \subset \cdots \subset X^{n-1} \subset X^n$ with $\cup X^n = X$.

Figure 9.13 shows a cell decomposition of a two-dimensional space with the one- and zero-dimensional skeletons.

Example 9.27   The open simplexes of a polyhedron $|K|$ are a cell decomposition of $|K|$. This means that the disjoint union of 0-, 1-, and 2-cells (for instance a two-dimensional polyhedron) is equal to $|K|$.

Definition 9.31 (Closure and Boundary of Cells). For every cell we have a closed cell $\bar{c}$ or the closure of c in X. The difference $\partial c = \bar{c} - c$ is the boundary of c.

Cell decomposition
of a 2-dimensional space

1-dimensional skeleton

0-dimensional skeleton

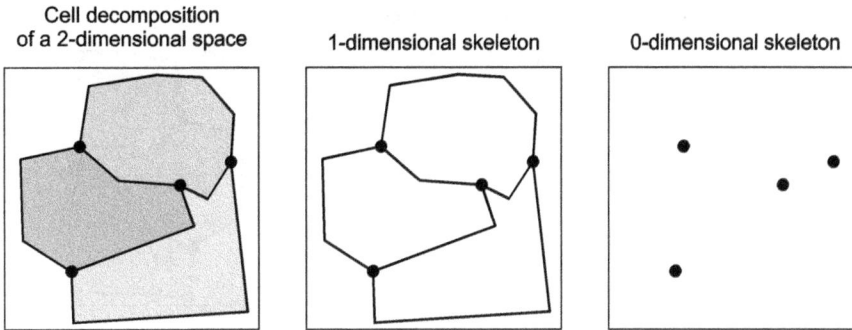

Figure 9.13   Cell decomposition and skeletons

It is important to note that the boundary of a cell in general is not the same as the boundary of a set. The boundary of a set is always defined with regard to an embedding space, whereas the boundary of a cell is depending on the dimension of the cell which is clearly determined.

Example 9.28   Consider a line segment L in $\mathbb{R}^3$. The point set topological boundary of L is defined as $\partial L = \overline{L} \cap \overline{\mathbb{R}^3 - L}$, which is all of $\overline{L}$. If, however, L is a 1-cell, then its boundary are the two end points.

Definition 9.32 (Cell Complex). A Hausdorff space X with a cell decomposition is a *cell complex* (or *CW complex*) if the following conditions are met:

(1) For every cell c there exists a continuous function $f: D^n \rightarrow X$ such that $f(S^{n-1}) \subset X^{n-1}$ and the open cell c is a homeomorphic image of the unit cell, i.e., $f(\mathring{D}^n) = c$.

(2) Every closed cell $\overline{c}$ is contained in a finite union of open cells.

(3) A subset $A \subseteq X$ is closed if and only if $A \cap \overline{c}$ is closed in X for every cell c.

A CW complex is n-dimensional if $X = X^n \neq X^{n-1}$. If a cell complex has a finite number of cells it is called a finite CW complex.

Condition (1) defines a function from the n-dimensional unit ball to the space X such that the open cell appears as a homeomorphic image of the unit cell and the $(n-1)$-sphere is continuously mapped to a subset of the X^{n-1}-space. In particular, we have $f : (D^n, S^{n-1}) \rightarrow (\overline{c}, \partial c)$ or $f(\mathring{D}^n \cup S^{n-1}) = c \cup \partial c$. The closed cell and the boundary are compact. Condition (2) is also called closure finite; condition (3) is the condition for the so-called weak topology.

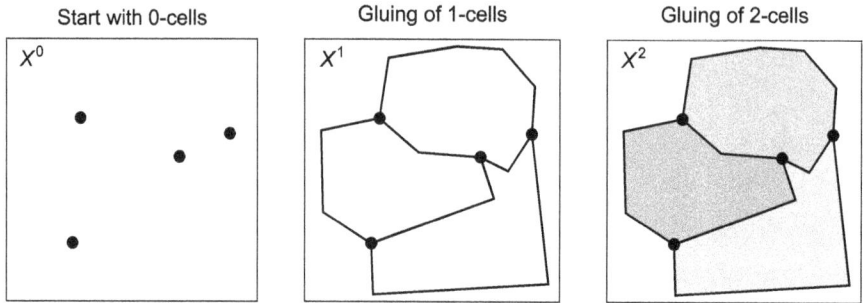

Figure 9.14   Construction of a CW complex

The following statements underline the differences between CW complexes and simplicial complexes:

(1) Cells of a CW complex need not be geometric simplexes.

(2) The closure of a n-cell need not be a n-ball and the boundary of a n-cell need not be a $(n-1)$-sphere.

(3) Not for every $k < n$ with n being the dimension of the CW complex there need to be cells of dimension k. However, every nonempty CW complex has at least one 0-cell.

(4) The closure $\bar{c}$ and boundary ∂c of a cell need not be the union of cells.

CW complexes can be easily constructed. Figure 9.14 illustrates the construction of a two-dimensional CW complex. We start with a discrete space X^0 (consisting of at least one 0-cell); we then glue 1-cells so that we get X^1, then we glue 2-cells which gives us X^2. We see that $X^0 \subset X^1 \subset X^2$.

9.5   Applications in GIS

The space used in GIS to represent spatial features is predominantly the two- or three-dimensional Euclidean space $\mathbb{R}^2$ or $\mathbb{R}^3$ equipped with the natural topology of open disks and balls. The Euclidean space is a metric space (therefore also a normal, regular, and Hausdorff space), second-countable (therefore also first-countable, separable, and Lindelöf), and connected. Closed and bounded subsets of a Euclidean space are compact (such as closed cells and simplexes).

Spatial data sets consisting of linear features (network) in $\mathbb{R}^2$ or $\mathbb{R}^3$ are one-dimensional CW complexes where the arcs are the 1-cells and the nodes are the zero-dimensional skeleton. Polygon feature data sets are two-dimensional CW complexes with the polygons as 2-cells and the bounding arcs and nodes as the one-dimensional skeleton.

9.5.1   Spatial Data Sets

To represent two-dimensional spatial features in a GIS we have two options: (i) to use a simplicial complex, i.e., to represent all spatial features as a set of simplexes with certain conditions (see Definition 9.28 of a simplicial complex), or (ii) to represent them as a cell complex by considering the cells being glued together in a proper way (see Definition 9.32 of a CW complex).

In the first case, we must represent all features by a set of triangles. On one hand, triangles are very simple structures and easy to handle; on the other hand, every polygon has to be approximated by a potentially large number of triangles which is often undesirable. An exception is a triangular irregular network (TIN) to represent a digital elevation model.

In the second case all features are cells glued together. This approach is much more suitable for general polygon features, because it avoids the use of triangles. In fact, every topologically structured data set in a GIS database is a digital representation of a two-dimensional cell complex.

Figure 9.15 shows a two-dimensional spatial data set as a cell complex embedded in $\mathbb{R}^2$. This complex consists of four 0-cells (1, 2, 3, 4), six 1-cells (a, b, c, d, e, f), and three 2-cells (A, B, C). The embedding Euclidean space $\mathbb{R}^2$ functions as the "world polygon" or "outside polygon" often denoted as W or O.

A data structure to represent this cell complex is the so-called arc-node structure. The 0-cells are the nodes and the 1-cells are the arcs between the nodes. Every arc has a start node and an end node, thereby defining an orientation of the arc.[8] The orientation is indicated by arrows in the figure. For every arc we note which polygon (2-cell) lies to the left and which one to the right of it viewed in the direction from start node to end node.

---

[8]The orientation of an arc is usually determined by the digitization or measuring process, i.e., a line is followed from the beginning (start-node) to the end (end-node).

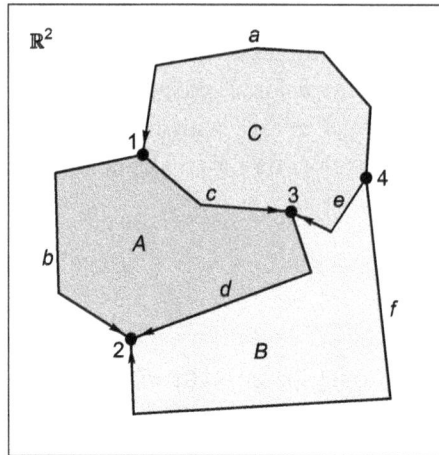

Figure 9.15   Two-dimensional spatial data set as cell complex

Networks, like road or river networks, are best modeled as a one-dimensional subset (or skeleton) of a cell complex. The topological relationships are then reduced to incidence relationships between edges (arcs) and nodes. Such a structure is also called a graph. Graph theory, although closely related to topology, has developed as an independent mathematical discipline. A special type of graph frequently used in GIS is a planar graph. It is completely embedded in the plane such that no two edges intersect except at nodes.

An implementation of the arc-node structure in a relational database needs one table for the arc-node relationships, one for the polygon attributes, and one for the vertices of the arcs. Table 9.2 shows the arc table of the arc-node structure of the cell complex in Figure 9.15.

Table 9.2   Arc table for the arc-node structure

Arc-id	Start-node	End-node	Left-polygon	Right-polygon
a	4	1	*C*	*W*
b	1	2	*A*	*W*
c	1	3	*C*	*A*
d	3	2	*B*	*A*
e	4	3	*B*	*C*
f	4	2	*W*	*B*

9.5.2 Topological Transformations

As we have seen, topology is the branch of mathematics that deals with properties of spaces that remain invariant (i.e., do not change) under topological mappings. Assume you have spatial features stored in a database using the arc-node structure. When you apply a transformation (such as a map projection) to the data set, the neighborhood relationships between A, B, and C remain, and the boundary lines have the same start and end nodes. The areas are still bounded by the same boundary lines, and only their shapes, areas, and the lengths of the perimeters have changed (Figure 9.16).

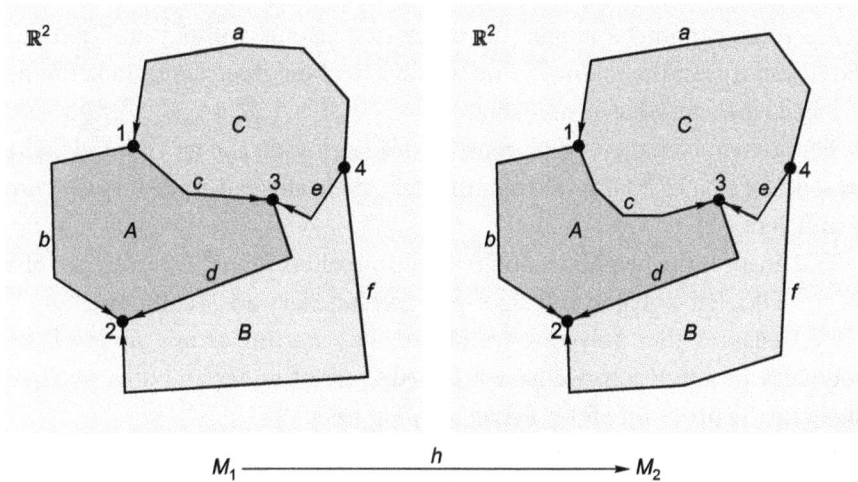

Figure 9.16   Topological mapping

Topologically speaking, we have applied a homeomorphism $h : M_1 \rightarrow M_2$ from the cell complex M_1 to the cell complex M_2. They are topologically equivalent.

9.5.3 Topological Consistency

A representation of a cell complex must be consistent, i.e., the topological properties must not be violated. If we can show that the following rules are satisfied for every element in the data set, we know that it is a topologically consistent two-dimensional configuration.

(TC1) Every 1-cell is bounded by two 0-cells (Every arc has a start-node and an end-node).

(TC2) For every 1-cell there are two 2-cells (For every arc there exist two adjacent polygons, the left- and right-polygon).

(TC3) Every 2-cell is bounded by a closed cycle of 0- and 1-cells (Every polygon has a closed boundary consisting of an alternating sequence of nodes and arcs).

(TC4) Every 0-cell is surrounded by a closed cycle of 1- and 2-cells (Around every node there exists an alternating closed sequence of arcs and polygons).

(TC5) Cells intersect only in 0-cells (If arcs intersect, they do so in nodes).

These rules cannot be applied to other dimensions without additions or modifications. In the following, we will discuss how these conditions can be checked when we have an arc table.

TC1 demands that every arc must have a start node and an end node. The presence of a NOT NULL value for the start-node and end-node for every arc is sufficient.

TC2 ensures the neighborhood relationship of polygons. The presence of a NOT NULL left-polygon and right-polygon for every arc is sufficient.

TC3 ensures that polygons are closed, i.e., starting at any node of the boundary of a polygon, we have a closed cycle of nodes and arcs. We will illustrate the procedure for polygon A in Figure 9.15.

Select all rows from the arc table where A appears either as right-polygon or left-polygon, as shown in Table 9.3.

Table 9.3 All rows from the arc table where A appears either as right-polygon or left-polygon

Arc-id	Start-node	End-node	Left-polygon	Right-polygon
a	4	1	C	W
b	1	2	A	W
c	1	3	C	A
d	3	2	B	A
e	4	3	B	C
f	4	2	W	B

Make sure that for all selected records A appears always as the left-polygon or always as the right-polygon. In our example, we want A always to be the

Table 9.4 Configuration after swapping for arc *b*

Arc-id	Start-node	End-node	Left-polygon	Right-polygon
a	4	1	*C*	*W*
b	2	1	*W*	*A*
c	1	3	*C*	*A*
d	3	2	*B*	*A*
e	4	3	*B*	*C*
f	4	2	*W*	*B*

right-polygon.[9] For those rows where A is not the right-polygon, we must swap left- and right-polygon. Of course, if we do that, we must also swap start- and end-node to maintain orientation. In our case, we must swap for arc *b*, which results in the configuration as shown in Table 9.4.

We now start at any start-node of the selected rows and chain through the nodes. In our example, let us start with arc *c* and node 1 (Figure 9.17). The end-node of *c* is 3. In the next step look up the record where 3 appears as the start-node and continue as before. When we return to the node where we started, the cycle is closed and the polygon boundary is closed. Otherwise, we have an inconsistency in the polygon boundary.

TC4 ensures the planarity of the cell complex near a node, i.e., for every node there must be an "umbrella" of a closed cycle of alternating 1-cells and 2-cells. We will illustrate a procedure for node 3 in Figure 9.15.

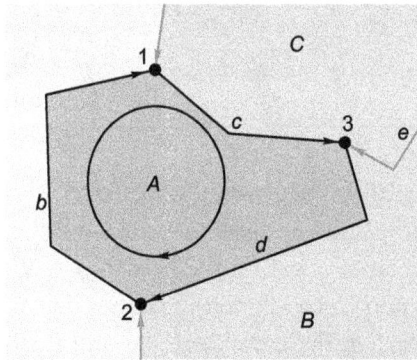

Figure 9.17 Closed polygon boundary check

---

[9] The choice could be based on the fact that for two out of three rows this condition is already fulfilled.

Select all rows from the arc table where 3 appears either as start-node or end-node, as shown in Table 9.5.

Table 9.5   All rows from the arc table where 3 appears either as start-node or end-node

Arc-id	Start-node	End-node	Left-polygon	Right-polygon
a	4	1	C	W
b	1	2	A	W
c	1	3	C	A
d	3	2	B	A
e	4	3	B	C
f	4	2	W	B

Make sure that for all selected records 3 appears always as the start-node or always as the End-node. In our example, we want 3 always to be the end-node. For those rows where 3 is not the End-node, we must swap start- and end-node. Of course, if we do that, we must also swap left- and right-polygon to maintain orientation. In our case, we must swap for arc *d*, which gives us Table 9.6.

Table 9.6   Configuration after swapping for arc *d*

Arc-id	Start-node	End-node	Left-polygon	Right-polygon
a	4	1	C	W
b	2	1	W	A
c	1	3	C	A
d	2	3	A	B
e	4	3	B	C
f	4	2	W	B

We now start at any left-polygon of the selected rows and chain through the polygons. In our example, let us start with arc *c* and left-polygon C (Figure 9.18). The right-polygon of *c* is A. In the next step look up the record where A appears as the left-polygon and continue as before. When we return to the polygon where we started, the cycle is closed and the "umbrella" is closed. Otherwise, we have an inconsistency in the node.

TC5 must be checked by calculating intersections of arcs and pointing out intersections at locations without nodes.

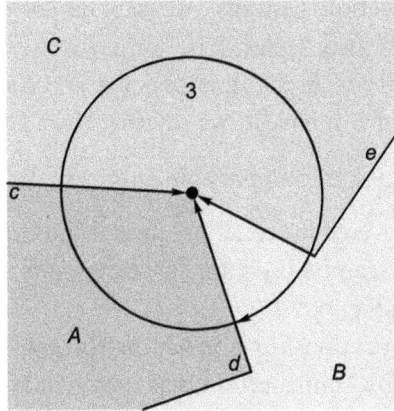

Figure 9.18   Node consistency check

9.5.4   Spatial Relations

Whereas relationships between simplexes or cells define consistency constraints for spatial data, we can use the topological properties of interior, boundary, and exterior to define relationships between spatial features. Since the properties of interior, boundary, and exterior do not change under topological mappings, we can investigate their possible relations between spatial features.

Let us assume two spatial regions A and B. Both have their respective boundaries, interiors, and exteriors. When we consider all possible combinations of intersections between the boundaries, the interiors, and the exteriors of A and B, we know that these will not change under any topological transformation. This can be put into a rectangular schema $I_9(A, B)$ which is called the 9-intersection, written as

$$I_9(A, B) = \begin{pmatrix} A^\circ \cap B^\circ & A^\circ \cap \partial B & A^\circ \cap B^- \\ \partial A \cap B^\circ & \partial A \cap \partial B & \partial A \cap B^- \\ A^- \cap B^\circ & A^- \cap \partial B & A^- \cap B^- \end{pmatrix}$$

A simpler version is the 4-intersection where only interior and boundary are considered. The 4-intersection can be written as:

$$I_4(A, B) = \begin{pmatrix} A^\circ \cap B^\circ & A^\circ \cap \partial B \\ \partial A \cap B^\circ & \partial A \cap \partial B \end{pmatrix}$$

From these intersection patterns, we can derive eight mutual spatial relationships between two regions. If, for instance, the boundary of A intersects the boundary of B, the interiors of A and B do not intersect, and the exteriors of A and B intersect, we say that A and B "meet", or $\boldsymbol{I}_{meet} = \begin{pmatrix} \varnothing & \varnothing \\ \varnothing & \neg\varnothing \end{pmatrix}$. Figure 9.19 shows all possible eight spatial relationships: disjoint, meet, equal, inside, covered by, contains, covers, and overlap.

These relationships can be used, for instance, in queries against a spatial database, as shown in Figure 9.19.

Let us assume the relationship A inside B. The graphic representation of the relationship and the 4-intersection is as shown in Figure 9.20.

$$\boldsymbol{I}_{inside} = \begin{pmatrix} \neg\varnothing & \varnothing \\ \neg\varnothing & \varnothing \end{pmatrix}$$

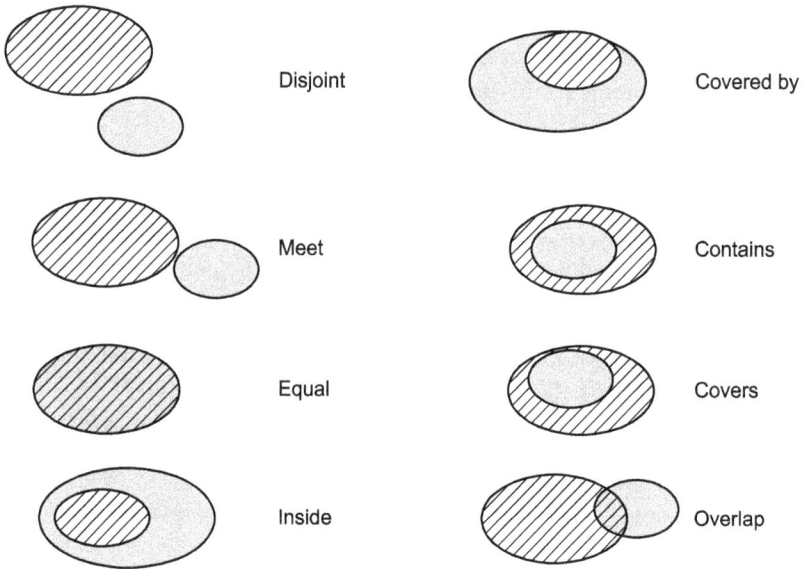

Figure 9.19   Spatial relationships between two simple regions based on the 9-intersection

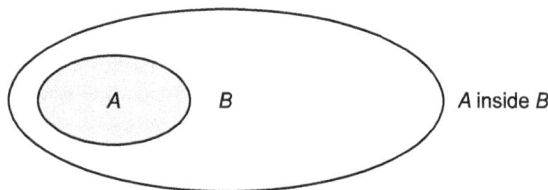

Figure 9.20   Relationship of A inside B in the 4-intersection model

9.6 Exercises

Exercise 9.1 A cell complex with 9 arcs, 6 nodes and 4 polygons is shown in Figure 9.21. Complete the arc-table and check the consistency of polygon *A* (closed boundary condition) and node 4 ("umbrella" condition).

Exercise 9.2 Let *A* and *B* be two regions. Write down the 4-intersection matrix for the following spatial relations: *A* covers *B*, *A* meets *B*, *A* contains *B*.

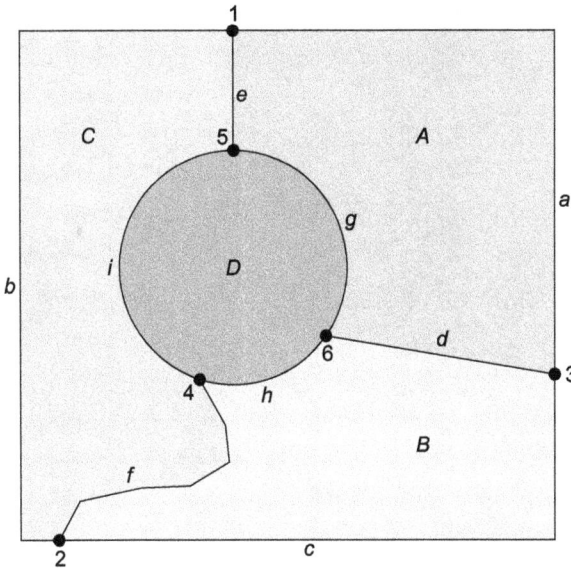

Figure 9.21 Cell complex

Chapter 10

Ordered Sets

One of the basic structures mathematical disciplines are built upon is order. A set is said to be (partially) ordered when an order relation is defined between its elements, which makes them comparable. The study of partially ordered sets and lattices (a special kind of ordered set) is covered by an extensive amount of mathematical literature. This theory has mainly been applied in computer science, such as in multiple inheritance or Boolean algebra.

In this chapter, we will introduce the basic principles of partially ordered sets and lattices and show how they can be applied to spatial features and their relationships with each other.

10.1 Posets

Definition 10.1 (Partially Ordered Set). Let P be a set. A binary relation $\leqslant$ on P such that for every $x, y, z \in P$:
(1) $x \leqslant x$ (reflexive)
(2) if $x \leqslant y$ and $y \leqslant x$, then $x = y$ (antisymmetric)
(3) if $x \leqslant y$ and $y \leqslant z$, then $x \leqslant z$ (transitive)

is called a *partial order* on P. A set P equipped with a reflexive, antisymmetric and transitive relation (order relation) $\leqslant$ is called a *partially ordered set* (or *poset*) and is written as $(P; \leqslant)$. Usually, we will write only P with the meaning "P is a poset".

For every partially ordered set P we can find a new poset, the dual of P, by defining that $x \leqslant y$ in the dual if and only if $y \leqslant x$ in P. Any statement about a partially ordered set can be turned into a statement of its dual by replacing $\leqslant$ with $\geqslant$ and vice versa.

Example 10.1   The natural numbers with the relation $\leqslant$ read as "less than or equal" are a poset.

Example 10.2   When we take the power set $\wp(X)$ of a set X, i.e., all subsets of X, then $\wp(X)$ is ordered by set inclusion and for every $A, B \in \wp(X)$ we define $A \leqslant B$ if and only if $A \subseteq B$.

Example 10.3   For spatial subdivisions A and B the order relation $A \leqslant B$ means that "A is contained in B" or dually, that "B contains A".

Any hierarchy is a poset with at most one element directly above any element. A special type of hierarchy—and therefore a more special type of poset—is the totally *ordered set* (or *chain*). This is a hierarchy in which at most one element is directly below any specific element, which means that every element can be compared with every other element in the set. The integer space is a typical example of a totally ordered set.

10.1.1   Order Diagrams

For every (finite) poset there exists a graphical representation, the diagram (or *Hasse diagram*) of the poset. To describe how to construct a diagram we need the idea of covering.

Definition 10.2 (Cover).   By "*A covers B*" (or "*B is covered by A*") in a poset P we mean that $B \leqslant A$ and there exists no $x \in P$ that $B < x < A$ and we write $A > - B$ or $B - < A$. In other words, A covers B means that A is immediately greater than B and there is no other element in between. The set of all elements that cover an element X is called the cover of X, written as X^-. Dually, the set of all elements that are covered by X is called the cocover of X and we write X_-.

A diagram of a poset P is drawn as a configuration of circles (representing the elements of P) and connecting lines (indicating the covering relation), where the circle for element A is drawn above the circle for element B, when A covers B. The circles are connected with straight lines. For a finite poset we obtain the diagram of the dual by turning it upside down. Figure 10.1 shows a poset and its corresponding diagram.

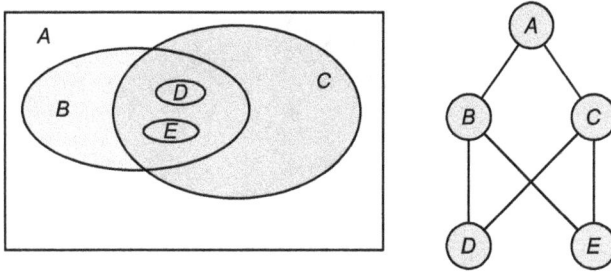

Figure 10.1   Poset and corresponding diagram

The circles and the connecting lines in a diagram can be viewed as vertices and edges of a graph. Since the order relation defines a direction for the edges, a Hasse diagram is a directed graph. There are no cycles in that graph, i.e., starting from a specific node and moving along the edges of the graph in the given direction, no node is visited twice. Such a graph is called a *directed acyclic graph* (or DAG). There are many algorithms for traversing directed acyclic graphs and for other related operations.

Definition 10.3 (Maximum and Minimum). Let P be a poset and $S \subseteq P$. An element $a \in S$ is the *greatest* (or *maximum*) *element* of S if $a \geqslant x$ for every $x \in S$ and we write $a = \max S$. The greatest element of P, if it exists, is called the *top element* of P. The *least* (or *minimum*) *element* of S, written as $\min S$, and the bottom element of P, if it exists, is defined by duality.

Example 10.4   In $\wp(X)$ we have X as the top element and the empty set as the bottom element.

Example 10.5   The natural numbers under their usual order have 1 as the bottom element but no top element.

10.1.2   Upper and Lower Bounds

Definition 10.4 (Upper Bound). Let P be a poset and $S \subseteq P$. An element $x \in P$ is an *upper bound* of S if $s \leqslant x$ for all $s \in S$. A *lower bound* is defined by duality. The set of all upper bounds of S is denoted by S^* (or "S *upper*") and the set of all lower bounds (or "S *lower*") is written as S_*; in other words we define $S^* = \{x \in P | \forall_{s \in S} s \leqslant x\}$ and $S_* = \{x \in P | \forall_{s \in S} s \geqslant x\}$.

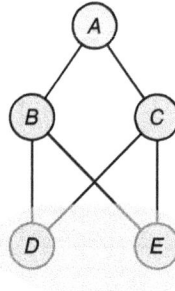

Figure 10.2   Lower bounds

If S^* has a least element, it is called *least upper bound* (l.u.b.), also *join* or *supremum*. By duality, if S_* has a largest element, it is called *greatest lower bound* (g.l.b.), *meet* or *infimum*. If a least upper bound or a greatest lower bound exists, it is always unique. For the least upper bound and the greatest lower bound of two elements x and y we write $\sup\{x, y\}$ or $x \vee y$ (read as "x *join* y") and $\inf\{x, y\}$ or $x \wedge y$ (read as "x *meet* y"), respectively. For a subset S we write $\vee S$ (the "*join of* S") or $\sup S$ and $\wedge S$ (the "*meet of* S") or $\inf S$.

There are cases when a greatest lower bound or a least upper bound does not exist. This may be the case because elements do not have common bounds or because a g.l.b. or l.u.b. does not exist. Take for example the two elements B and C of Figure 10.1. The set of their lower bounds is $\{D, E\}$. However, none of the lower bounds is greater than the other, they are not comparable. Therefore, there is no greatest lower bound for the subset $\{B, C\}$ (Figure 10.2).

10.2   Lattices

In the previous section, we have seen that in the general case of a partially ordered set we cannot expect that join and meet always exist. Therefore, a more specific order structure is needed.

Definition 10.5 (Lattice). A lattice L is a poset in which every pair of elements has a least upper bound and a greatest lower bound. A lattice is called *complete*, when meet and join exist for every subset of the poset.[1]

---

[1]Note that the difference between a lattice and a complete lattice is in the existence of the meet and join for every *pair* of elements (lattice) or every *subset* of elements (complete lattice).

If L is a lattice then $\wedge$ and $\vee$ are binary operations on L and we have an algebraic structure $< L, \wedge, \vee >$ with $\wedge$ and $\vee$ satisfying the following conditions for all $a, b, c \in L$:

(1) $a \vee (b \vee c) = (a \vee b) \vee c, a \wedge (b \wedge c) = (a \wedge b) \wedge c$ (associative laws).
(2) $a \vee b = b \vee a, a \wedge b = b \wedge a$ (commutative laws).
(3) $a \vee a = a, a \wedge a = a$ (idempotency laws).
(4) $a \vee (a \wedge b) = a, a \wedge (a \vee b) = a$ (absorption laws).

Every set L with two binary operations satisfying conditions (1) to (4) is a lattice. We see that a lattice can be viewed as either an order structure or an algebraic structure. Many theories, e.g., Boolean algebras, rely heavily on the algebraic properties of lattices.

The order relation $\leqslant$ is related to the algebraic operations of $\wedge$ and $\vee$ by the following statement: Let L be a lattice and let x and y be elements of L. Then $x \leqslant y$ is equivalent to each of the conditions: $x \wedge y = x$ and $x \vee y = y$.

It can be proven that every finite lattice is complete. This is an important result because it means that whenever we have a lattice with a finite number of elements we can always find least upper bounds and greatest lower bounds for every subset of the lattice.

Example 10.6 Every chain is a lattice in which $x \vee y = \max\{x, y\}$ and $x \wedge y = \min\{x, y\}$.

Therefore, the natural numbers, integers, rational and real numbers are all lattices under their usual orders. None of them is a complete lattice. To show this let us take any of these sets and determine the supremum of the set itself. Since there is no greatest number in any of these sets, the set of the upper bounds is empty and a least upper bound does not exist.

Example 10.7 The power set $\wp(X)$ of any set X is a complete lattice where meet and join are defined as $\wedge\{A_i | i \in I\} = \bigcap_{i \in I} A_i$ and $\vee\{A_i | i \in I\} = \bigcup_{i \in I} A_i$, respectively.

If a subset $S \in \wp(X)$ is closed under finite unions and intersections, it is called a *lattice of sets*. It is called a *complete lattice* of sets if it is closed under arbitrary unions and intersections. Meet and join are then defined as set intersection and set union.

If L is a complete lattice then the following is true for every $S, T \subseteq L$:
(1) $\forall s \in S, s \leqslant \vee S$ and $s \geqslant \wedge S$.
(2) Let $x \in L$. Then $x \leqslant \wedge S$ if and only if $x \leqslant s$ for all $s \in S$.

(3) Let $x \in L$. Then $x \geqslant \vee S$ if and only if $x \geqslant s$ for all $s \in S$.
(4) $\vee S \leqslant \wedge T$ if and only if $s \leqslant t$ for all $s \in S$ and all $t \in T$.
(5) If $S \subseteq T$, then $\vee S \leqslant \vee T$ and $\wedge S \geqslant \wedge T$.
(6) $\vee(S \cup T) = (\vee S) \vee (\vee T)$ and $\wedge(S \cup T) = (\wedge S) \wedge (\wedge T)$.

10.3   Normal Completion

Not every poset is a lattice, because posets exist in which not all subsets have greatest lower bounds and least upper bounds. For example, the subset $\{B, C\}$ of the order in Figure 10.2 has no greatest lower bound. It is, however, possible, to add elements to a poset to create a lattice. This is in fact possible for all posets.

It is even more interesting to find the smallest number of elements necessary to add to a poset to create a lattice. In other words, we want to build the minimal containing lattice of a poset. The method for doing this is called *normal completion.*

In order to define the normal completion, we need the concept of a closure operator, which is defined as follows.

Definition 10.6 (Closure). Let X be a set. A map $C : \wp(X) \rightarrow \wp(X)$ is called a closure operator on X if, for all $A, B \subseteq X$:
(1) $A \subseteq C(A)$.
(2) If $A \subseteq B$, then $C(A) \subseteq C(B)$.
(3) $C(C(A)) = C(A)$.

A subset A of X is called *closed* if $C(A) = A$.

The following theorem summarizes the important facts about the normal completion of posets. It even gives us a procedure for building the normal completion lattice.

Let P be a poset and $(A^*)_*$ be the set of the lower bounds of the upper bounds of a subset A of P. Then
(1) $C(A) = (A^*)_*$ defines a closure operator on P.
(2) The family $\mathrm{DM}(P) = \{A \subseteq P | (A^*)_* = A\}$ is a complete lattice (the *Dedekind-MacNeille completion*, or *normal completion* or *completion by cuts* of P), when ordered by inclusion, in which $\wedge\{A_i | i \in I\} = \cap_{i \in I} A_i$ and $\vee\{A_i | i \in I\} = C(\cup_{i \in I} A_i)$.

(3) The map $\varphi\colon P \to DM(P)$ defined by $\varphi(x) = (x^*)_*$ for all $x \in P$ is an *order-embedding*, i.e., it is order-preserving and injective. In fact, φ can be defined as $\varphi(x) = x_* = \{y \in P | y \leqslant x\}$, because $(x^*)_* = x$ for all $x \in P$. $DM(P)$ is a completion of P via φ and all greatest lower bounds and least upper bounds which exist in P are preserved. This means, if $A \subseteq P$ and $\vee A$ exists in P, then $\varphi(\vee A) = \vee \varphi(A)$, and $\varphi(\wedge A) = \wedge \varphi(A)$.

(4) $DM(P)$ is the smallest lattice in which P can be embedded in the sense that if L is any other lattice such that $P \subseteq L$, we have $P \subseteq DM(P) \subseteq L$.

By calculating the normal completion, we have to look at all subsets of the poset P. For practical applications, this is rather inefficient, because every set with n elements has 2^n subsets.

The theorem above has a simple corollary, which yields two important properties of the normal completion lattice:

(1) If L is a lattice, then $L = DM(L)$.

(2) For all posets P we have $DM(P) = DM(DM(P))$.

Firstly, the corollary tells us that whenever the poset is already a lattice, the normal completion does not add anything to the lattice. It leaves the lattice unchanged. Secondly, it follows from the idempotency of a closure operator that applying the completion more than once does not increase the number of elements added to the completion lattice, i.e., the number of elements in the completion lattice is bounded by 2^n for n elements in the poset.

10.3.1   Special Elements

Let P be a poset and S be a subset of the poset. We had defined upper bounds and lower bounds for the subset S. The set of all upper bounds of S is denoted as S^* and the set of all lower bounds of S is denoted as S_*. For the normal completion lattice we need to identify all $(S^*)_*$ for all subsets of P.

If there exists a greatest element in P it is called top element and written as $\top$; if there is a least element in P it is called bottom element and written as $\bot$.

There are two cases that require special attention: when $S = P$ and when $S = \emptyset$. First, let us investigate the case when $S = P$. If P has a top element, then $P^* = \{\top\}$ and $\sup P = \top$. When P has no top element, then $P^* = \emptyset$

and there is no supremum of P. By duality, if P has a bottom element, then $P_* = \{\bot\}$ and inf $P = \bot$. If P does not have a bottom element, then $P_* = \emptyset$ and the infimum does not exist.

Now, let us assume that $S = \emptyset$, i.e., S is the empty subset of P. Then (vacuously) for all $s \in S$ we have that $s \leqslant x$ for every element $x \in P$. Thus, $\emptyset^* = P$ and sup P exists, if and only if P has a bottom element; i.e., then we have sup $P = \bot$. Dually, $\emptyset_* = P$ (because again we have vacuously that for all $s \in S = \emptyset$, $s \geqslant x$ for every $x \in P$) and inf $P = T$ whenever P has a top element. Table 10.1 summarizes the result.

Table 10.1 Special elements and the closure operator in the normal completion

Subset	S^*		S_*	
	Top element exists	No top element	Bottom element exists	No bottom element
P	$\{T\}$	$\emptyset$	$\{\bot\}$	$\emptyset$
$\emptyset$	P	P	P	P

From Table 10.1, we can derive the following sets:

$$(P^*)_* = P$$

$$(\emptyset^*)_* = \begin{cases} \{\bot\} & \text{if } P \text{ has a bottom element} \\ \emptyset & \text{otherwise} \end{cases}$$

10.3.2 Normal Completion Algorithm

The algorithm for the normal completion can be written as follows:

(1) Determine all subsets, i.e., the power set $\wp(P)$ of the poset P.

(2) For every subset $S \in \wp(P)$ determine $(S^*)_*$.

(3) Arrange all $(S^*)_*$ to a poset where $\subseteq$ (subset) is the order relation.

(4) Identify every element $a \in P$ of the original poset with its corresponding $(a^*)_*$ in the new poset.

(5) Assign suitable symbols to the remaining elements of the new poset.

(6) The resulting poset is the normal completion lattice of P.

To illustrate how this works we take the poset of Figure 10.1 and build the normal completion lattice according to the algorithm. First, we determine all subsets of the poset $\{A, B, C, D, E\}$. This results in 32 sets. For every subset S we must then compute $(S^*)_*$. The result is given in Table 10.2.

The resulting sets are $\{A, B, C, D, E\}$, $\{B, D, E\}$, $\{C, D, E\}$, $\{D, E\}$, $\{D\}$, $\{E\}$, and $\emptyset$. When we arrange them in a poset according to the subset relation, we get the normal completion lattice (Figure 10.3).[2]

Finally, we identify the original poset elements with their corresponding lattice elements and denote the newly created elements with X and $\{\}$. Figure 10.4 shows the normal completion of the poset. We see that two new elements were added to the poset to form a lattice.

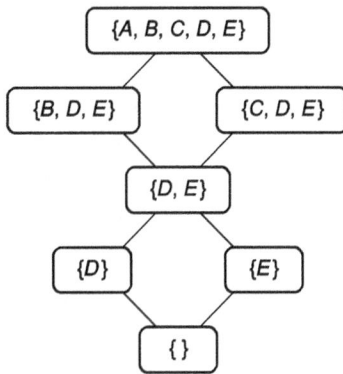

Figure 10.3   Normal completion lattice

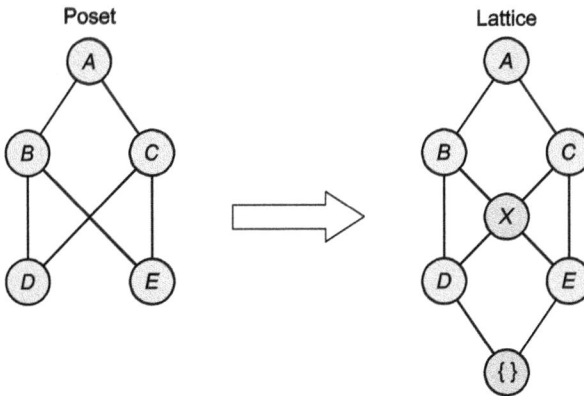

Figure 10.4   Normal completion

---

[2]Note that we may use either $\{\}$ or $\emptyset$ to denote the empty set.

Table 10.2　Normal completion

	S	S^*	$(S^*)_*$
(1)	$\emptyset$	$\{A, B, C, D, E\}$	$\emptyset$
(2)	$\{A\}$	$\{A\}$	$\{A, B, C, D, E\}$
(3)	$\{B\}$	$\{A, B\}$	$\{B, D, E\}$
(4)	$\{C\}$	$\{A, C\}$	$\{C, D, E\}$
(5)	$\{D\}$	$\{A, B, C, D\}$	$\{D\}$
(6)	$\{E\}$	$\{A, B, C, E\}$	$\{E\}$
(7)	$\{A, B\}$	$\{A\}$	$\{A, B, C, D, E\}$
(8)	$\{A, C\}$	$\{A\}$	$\{A, B, C, D, E\}$
(9)	$\{A, D\}$	$\{A\}$	$\{A, B, C, D, E\}$
(10)	$\{A, E\}$	$\{A\}$	$\{A, B, C, D, E\}$
(11)	$\{B, C\}$	$\{A\}$	$\{A, B, C, D, E\}$
(12)	$\{B, D\}$	$\{A, B\}$	$\{B, D, E\}$
(13)	$\{B, E\}$	$\{A, B\}$	$\{B, D, E\}$
(14)	$\{C, D\}$	$\{A, C\}$	$\{C, D, E\}$
(15)	$\{C, E\}$	$\{A, C\}$	$\{C, D, E\}$
(16)	$\{D, E\}$	$\{A, B, C\}$	$\{D, E\}$
(17)	$\{A, B, C\}$	$\{A\}$	$\{A, B, C, D, E\}$
(18)	$\{A, B, D\}$	$\{A\}$	$\{A, B, C, D, E\}$
(19)	$\{A, B, E\}$	$\{A\}$	$\{A, B, C, D, E\}$
(20)	$\{A, C, D\}$	$\{A\}$	$\{A, B, C, D, E\}$
(21)	$\{A, C, E\}$	$\{A\}$	$\{A, B, C, D, E\}$
(22)	$\{A, D, E\}$	$\{A\}$	$\{A, B, C, D, E\}$
(23)	$\{B, C, D\}$	$\{A\}$	$\{A, B, C, D, E\}$
(24)	$\{B, C, E\}$	$\{A\}$	$\{A, B, C, D, E\}$
(25)	$\{B, D, E\}$	$\{A, B\}$	$\{B, D, E\}$
(26)	$\{C, D, E\}$	$\{A, C\}$	$\{C, D, E\}$
(27)	$\{A, B, C, D\}$	$\{A\}$	$\{A, B, C, D, E\}$
(28)	$\{A, B, C, E\}$	$\{A\}$	$\{A, B, C, D, E\}$
(29)	$\{A, B, D, E\}$	$\{A\}$	$\{A, B, C, D, E\}$
(30)	$\{A, C, D, E\}$	$\{A\}$	$\{A, B, C, D, E\}$
(31)	$\{B, C, D, E\}$	$\{A\}$	$\{A, B, C, D, E\}$
(32)	$\{A, B, C, D, E\}$	$\{A\}$	$\{A, B, C, D, E\}$

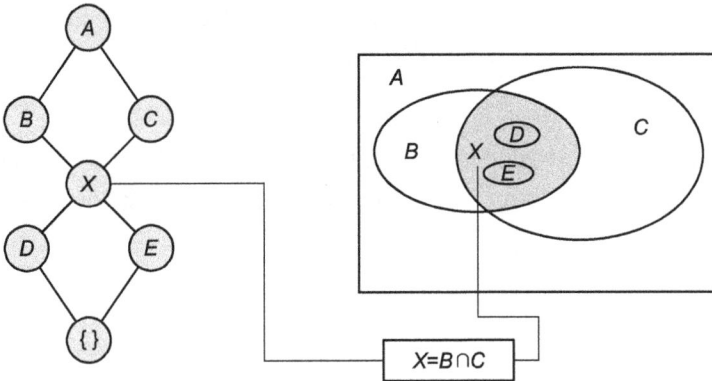

Figure 10.5   Geometric interpretation of new lattice elements

The new elements can be interpreted in a geometric way as shown in Figure 10.5. Element X can be interpreted as the intersection of B and C.

10.4   Applications in GIS

The intuitive interpretation of order relations as "is contained in" or, dually, as "contains" can be used for relationships among spatial features such as polygons, lines and points. The structure of a poset accommodates both strict hierarchies (every object has exactly one parent object) and relationships where one object possesses more than one parent object.

Examples of hierarchies are administrative subdivisions where for instance every county belongs to exactly one state, and every state belongs to exactly one country. General posets can be used to represent situations where one object belongs to several parents, such as agricultural production zones that may be part of several municipalities, or regions that are composed of several unconnected polygons such as the Hawaiian Islands.

10.5   Exercises

Exercise 10.1   From the poset in Figure 10.1 determine the upper bounds of (a) $\{D\}$, (b) $\{D,C\}$, (c) $\{A\}$.

Exercise 10.2   From the poset in Figure 10.1 determine the greatest lower bounds of (a) $\{B, D\}$, (b) $\{A\}$, (c) $\{A, B, C\}$.

Exercise 10.3   The following relationships are given for the four regions A, B, C, and D: C is contained in A and D is contained in B. Draw the poset for the four regions, and compute and draw the normal completion lattice.

Chapter 11

Graph Theory

The origin of graph theory lies in the investigation of topological problems given by a set of points and the connections between them. Today, graph theory is a branch of mathematics in its own right dealing with problems that can be represented by a collection of vertices and connecting edges.

This chapter deals with the basic principles of graphs, their representation and ways to traverse them.

11.1 Introducing Graphs

Generally, the origin of graph theory is attributed to the Swiss mathematician Leonhard Euler who published a paper in 1736 on what is now commonly known as the Königsberg bridge problem. Figure 11.1 shows a sketch of the seven bridges across the river Pregel in Königsberg (which is today's Kaliningrad). The problem is to determine whether it is possible to make a circular walk through Königsberg by starting at a river bank and crossing every bridge exactly once.

Euler solved the problem by abstracting the island and river banks to points and representing the bridges by lines connecting these points. In the figure they are represented by dots and connecting lines.

Figure 11.2 shows these points (*vertices*) and lines (*edges*) in a schematic way with the vertices numbered v_1 to v_4, and the edges e_1 to e_7. Such a configuration is called a *graph*. Starting from an arbitrary vertex we find after some tries that such a circular walk is impossible.[1]

---

[1] We will see later that the problem is to find an Eulerian circuit in the graph and that there is a theorem stating when such a circuit exists.

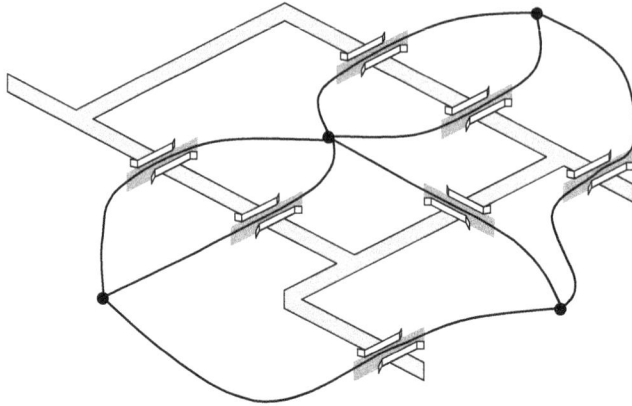

Figure 11.1   The seven bridges of Königsberg

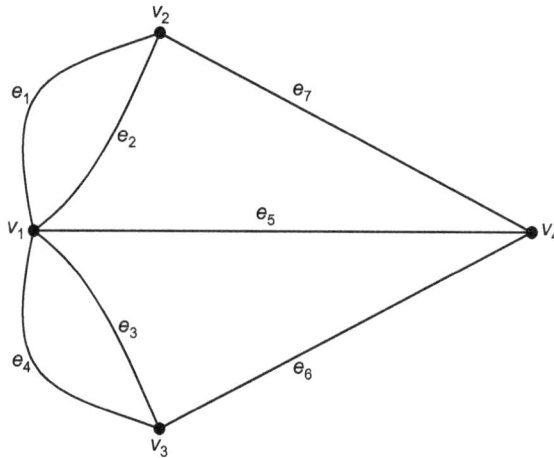

Figure 11.2   Graph of the Königsberg bridge problem

11.1.1   Basic Concepts

Definition 11.1 (Graph). Given a nonempty set $V = \{v_1, v_2, \ldots, v_n\}$, the *vertex-set*, a set $E = \{e_1, e_2, \ldots, e_m\}$, *the edge-set*, and a function $g \colon E \to V \times V$, the *incidence map*, which assigns to every element of E a pair of elements of $V(v_i, v_j)$. We call the triple $G = (V, E, g)$ a *graph*.

The elements of V are called *points* (or *vertices*), and the elements of E are called *edges*. For an edge $e = (v_i, v_j)$ vertices v_i and v_j are called *end points* of e; we say that e is *incident with* v_i and v_j, and that v_i *is adjacent to* v_j. If $g(e) = (v, v)$, we call e a *loop*. If $g(e_i) = g(e_j)$, we call e_i and e_j *parallel* edges.

Here, we will deal only with finite graphs, i.e., the number of vertices and the number of edges are both finite. When there is no confusion possible we will denote a graph with $G = (V, E)$ for short.

Example 11.1 The graph for the Königsberg bridge problem in Figure 11.2 can be written as $G = (V, E, g)$ with the vertex-set $V = \{v_1, v_2, v_3, v_4\}$, the edge-set $E = \{e_1, e_2, e_3, e_4, e_5, e_6, e_7\}$ and the incidence map defined as $g(e_1) = (v_1, v_2)$, $g(e_2) = (v_1, v_2)$, $g(e_3) = (v_1, v_3)$, $g(e_4) = (v_1, v_3)$, $g(e_5) = (v_1, v_4)$, $g(e_6) = (v_3, v_4)$, $g(e_7) = (v_2, v_4)$. Edges e_1, e_2 and e_3, e_4 are parallel. The graph contains no loop.

A graph without loops and parallel edges is called a *simple graph*. A graph with loops and parallel edges is sometimes called a *multigraph*. The number of edges incident with a vertex v is called the *degree* of v and is written as $d(v)$. A vertex v with $d(v) = 0$ is called an *isolated* vertex.

Example 11.2 In example 11.1, the degree of vertex v_1 is 5, and the degree of vertices v_2, v_3 and v_4 is three.

A graph is called *complete* if there exists an edge for every pair of distinct vertices. For n vertices the complete graph is denoted by K_n. We call a graph *regular* when every vertex has the same degree. If the degree is k then the graph is k-regular. A complete graph K_n is $(n-1)$-regular, i.e., every vertex in a complete graph K_n has degree $(n-1)$. Figure 11.3 illustrates some complete graphs.

Two graphs $G_1 = (V_1, E_1)$ and $G_2 = (V_2, E_2)$ are *isomorphic* if there is a bijective mapping $i: V_1 \to V_2$ preserving the incidence relationships, i.e., for every $v_1, v_2 \in V_1$ we have $(v_1, v_2) \in E_1$ implies $i(v_1, v_2) \in E_2$. Isomorphic graphs have the same structure, although they might look quite different at a first glance. Figure 11.4 shows two isomorphic graphs.

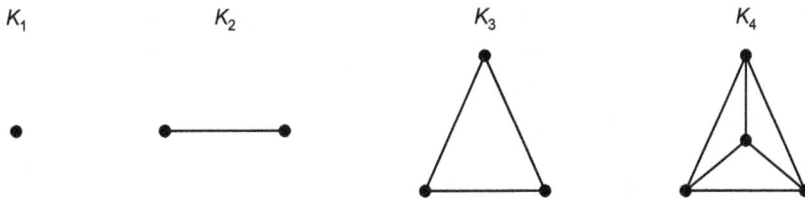

Figure 11.3   Complete graphs

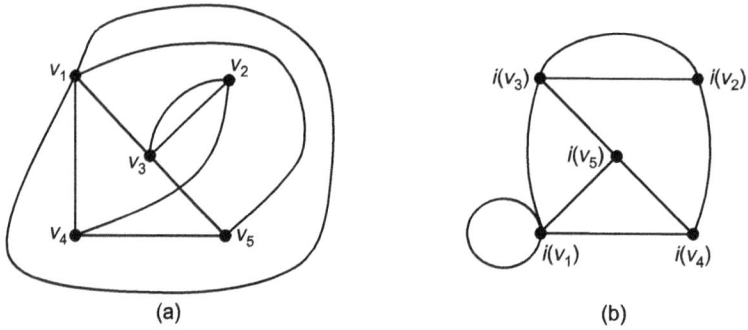

Figure 11.4   Isomorphic graphs

If we remove a number of edges or vertices from a graph G we obtain a *subgraph $S \subseteq G$*. The removal of a vertex implies that all edges incident with it must also be removed. However, if we remove an edge the vertices remain. The result could be some isolated vertices. A subset of vertices $V' \subseteq V$ and edges with both end-points in V' is called a subgraph *induced* by V'.

11.1.2   Path, Circuit, Connectivity

For many applications we need to traverse a graph, sometimes in a particular way. A path from v_1 to v_n is a sequence of alternating vertices and edges $P = v_1, e_1, v_2, e_2, \ldots, e_{n-1}, v_n$ such that for $1 \leqslant i < n$, e_i is incident with v_i and v_{i+1}. For a simple graph it is sufficient to list only the vertices in a path. If $v_1 = v_n$ then the path is called a *cycle* or *circuit*. A path is called a *simple path* if every vertex is visited only once. In a *simple circuit* every vertex appears once except that $v_1 = v_n$. The length of a path or a circuit is the number of edges it contains.

Example 11.3   In the graph of Figure 11.2, $P = v_1, e_1, v_2, e_7, v_4$ is a simple path from v_1 to v_4. $C = v_4, e_5, v_1, e_4, v_3, e_3, v_1, e_5, v_4$ is circuit. Note that it is not a simple circuit, because vertex v_1 is visited twice.

When we assign a number (weight) to each edge of a graph we get a *weighted* graph. In many applications such a weight is being used to represent the length of an edge as distance or travel time. This must not be confused with the length of a path as defined above.

Two vertices v_i and v_j are connected if there exists a path from v_i to v_j. Every vertex is connected to itself. A subgraph induced by a set of connected vertices

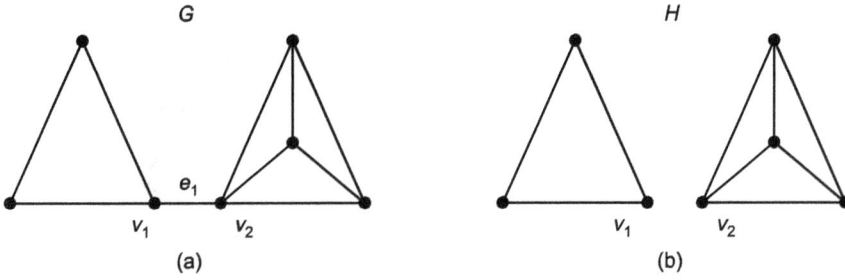

Figure 11.5   Connected (G) and disconnected (H) graphs

is called a *component* of a graph. A graph with only one component is called *connected*, otherwise it is *disconnected*. If the removal of a vertex v would disconnect the graph, then v is called an *articulation point*. A *block* is a graph without any articulation point. If the removal of an edge e would disconnect the graph then this edge is called a *cut-edge*.

Figure 11.5 shows an example of a connected graph and a disconnected graph. Graph H has two components. The vertices $v_1, v_2 \in G$ are articulation points; $e_1 \in G$ is a cut-edge.

11.2   Important Classes of Graphs

11.2.1   Directed Graph

If for every edge we assign one of its vertices as start point the graph becomes a *directed* graph (or *digraph*). We draw the edges of a digraph with arrows indicating their directions. A directed graph without cycles is called a *directed acyclic graph* (or DAG). DAGs play an important role in the representation of partially ordered sets. Digraphs are used to represent transportation or flow problems. Figure 11.6 shows two directed graphs, where Graph G contains a cycle and H is a DAG.

In a digraph an edge (v_i, v_j) is said to be *incident from* v_i and *incident to* v_j. The number of edges incident from a vertex v is called the *out-degree* $d^+(v)$, and the number of edges incident to v is the *indegree* $d^-(v)$. A digraph is *symmetric* if for every edge (v_i, v_j) there is also an edge (v_j, v_i). A digraph is *balanced* if for every vertex v the out-degree is equal to the in-degree, i.e., $d^+(v) = d^-(v)$.

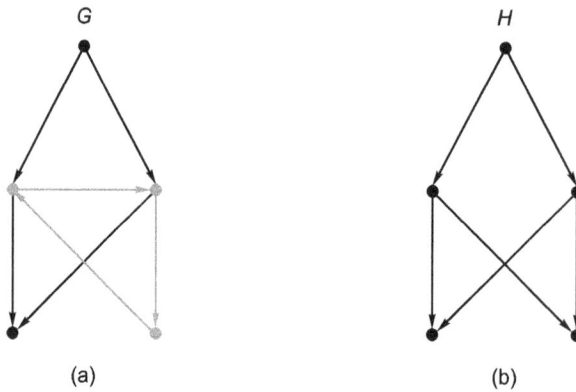

Figure 11.6   Directed graphs

11.2.2   Planar Graph

An important class of graphs is the planar graphs. A graph is *planar* if it can be drawn on a plane surface without intersecting edges.[2] Such a representation divides the plane into connected regions (or *faces*). The faces are bound by edges of the graph. If one face encloses the graph, this face is often called the exterior face.

A planar graph in the real drawing plane corresponds to a two-dimensional cell complex, where the vertices correspond to the 0-cells, the edges to the 1-cells, and the faces to the 2-cells. Clearly, this cannot be extended to higher dimensions.

Figure 11.7 shows a planar graph. This graph has four vertices $V = \{v_1, v_2, v_3, v_4\}$, six edges $E = \{e_1, e_2, e_3, e_4, e_5, e_6\}$, and four faces $F = \{f_1, f_2, f_3, f_4\}$. Face f_4 is the exterior face.

Euler's formula connects the number of vertices, edges and faces. It states that for every connected planar graph with n vertices, e edges, and f faces we have

$$n - e + f = 2$$

If we do not count the exterior face the formula changes to

$$n - e + f = 1$$

---

[2]This class of graphs plays an important role in the structuring of two-dimensional spatial data sets for GIS.

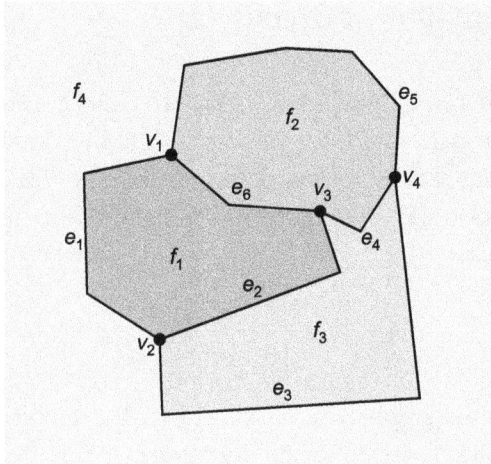

Figure 11.7   Planar graph

Example 11.4   For the planar graph in Figure 11.7 with four vertices, six edges, and four faces, we have $4 - 6 + 4 = 2$.

For every planar graph G, we can construct a graph G^* whose vertices are the regions of G; the edges represent the adjacency of faces, i.e., there is an edge connecting two vertices of G^* if the two corresponding faces of G are adjacent. The edge is drawn crossing the bounding edge of the faces in G. Such a graph is called the *dual* graph. It is again planar. Figure 11.8 shows the planar dual graph of the graph in Figure 11.7.

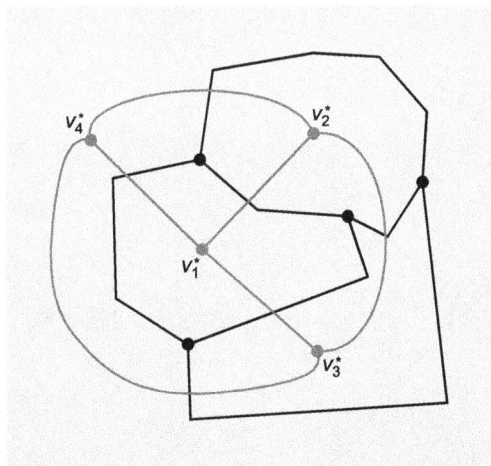

Figure 11.8   Planar dual graph

11.3   Representation of Graphs

For many computational purposes we need efficient data structures and algorithms to represent and traverse graphs. The best-known structures to represent a graph are adjacency matrices and adjacency lists.

Given a graph $G = (V, E)$ with n vertices an adjacency matrix is an $n \times n$ matrix A, such that:

$$A(i, j) = \begin{cases} 1 & \text{if } (i, j) \in E \\ 0 & \text{otherwise} \end{cases}$$

For an undirected graph $A(i, j) = A(j, i)$. For a digraph A is usually asymmetric. Figure 11.9 shows an undirected graph G_1 and a directed graph G_2.

If we sort the columns and rows from v_1 to v_5 the adjacency matrices for G_1 and G_2 are written as:

$$A(G_1) = \begin{pmatrix} 0 & 1 & 1 & 0 & 0 \\ 1 & 0 & 0 & 1 & 1 \\ 1 & 0 & 0 & 1 & 1 \\ 0 & 1 & 1 & 0 & 0 \\ 0 & 1 & 1 & 0 & 0 \end{pmatrix} \quad A(G_2) = \begin{pmatrix} 0 & 1 & 1 & 0 & 0 \\ 0 & 0 & 0 & 1 & 1 \\ 0 & 0 & 0 & 1 & 1 \\ 0 & 0 & 0 & 0 & 0 \\ 0 & 0 & 0 & 0 & 0 \end{pmatrix}$$

An adjacency list L shows for every vertex the vertices adjacent to it. The adjacency lists for the graphs in Figure 11.9 are:

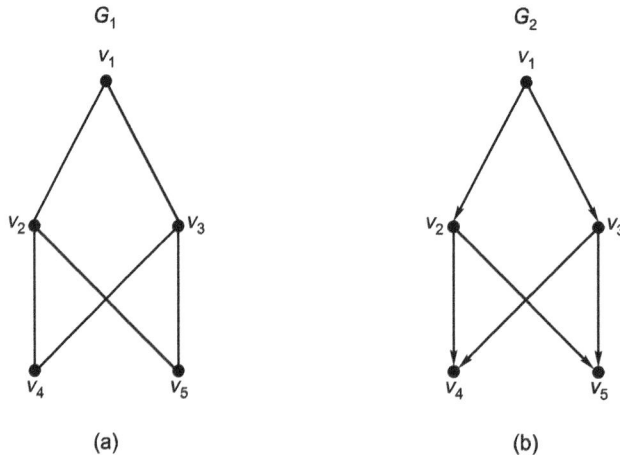

Figure 11.9   Undirected and directed graphs

$$L(G_1): \begin{bmatrix} v_1: & v_2, v_3 \\ v_2: & v_1, v_4, v_5 \\ v_3: & v_1, v_4, v_5 \\ v_4: & v_2, v_3 \\ v_5: & v_2, v_3 \end{bmatrix} \quad L(G_2): \begin{bmatrix} v_1: & v_2, v_3 \\ v_2: & v_4, v_5 \\ v_3: & v_4, v_5 \\ v_4: & - \\ v_5: & - \end{bmatrix}$$

We see easily that the storage requirement for adjacency matrices is usually higher than adjacency lists.

For directed acyclic graphs it is often more convenient to represent also the transitive relationships in the graph. This means that if we have $(v_i, v_j) \in E$ and $(v_j, v_k) \in E$ then we also show the relationship (v_i, v_k) in the matrix or list. The relationships (v_i, v_i) are trivially contained. This is called the *transitive closure* of the graph. For the directed acyclic graph G_2 in Figure 11.9, we represent the transitive closure as:

$$A(G_2) = \begin{pmatrix} 1 & 1 & 1 & 1 & 1 \\ 0 & 1 & 0 & 1 & 1 \\ 0 & 0 & 1 & 1 & 1 \\ 0 & 0 & 0 & 1 & 0 \\ 0 & 0 & 0 & 0 & 1 \end{pmatrix} \quad L(G_2): \begin{bmatrix} v_1: & v_2, v_3, v_4, v_5 \\ v_2: & v_4, v_5 \\ v_3: & v_4, v_5 \\ v_4: & - \\ v_5: & - \end{bmatrix}$$

Based on the representation of a graph by a matrix or a list, efficient algorithms can be formulated to traverse the graph. Traversal means that from a given start point all other vertices of the graph are visited. One of the best-known graph traversal algorithms is the *depth-first-search* (DFS). It works in the following way:

Step 1 Starting from a given vertex v visit an adjacent vertex that has not yet been visited.

Step 2 If no such vertex can be found then return to the vertex visited just before v and repeat Step 1.

11.4 Eulerian and Hamiltonian Tours, Shortest Path Problem

As mentioned above we often want to traverse a graph in a particular manner. When we do not distinguish between path and cycle we will talk about a *tour*.

A tour through a graph in which every edge is traversed exactly once is called an *Eulerian* tour. If we traverse the graph visiting each vertex exactly once we call this a *Hamiltonian* tour. The shortest path problem is about finding the shortest path between two given vertices in a graph.

11.4.1   Eulerian Graphs

An *Eulerian graph* is an undirected graph or digraph containing an Eulerian circuit. The following statements can be proven for undirected graphs and digraphs:

- An undirected graph contains an Eulerian circuit if and only if it is connected and the number of vertices with odd degree is 0.
- An undirected graph contains an Eulerian path if and only if it is connected and the number of vertices with odd degree is 2 (denoted with v_1 and v_2).
- A digraph contains an Eulerian circuit if and only if it is connected and balanced.
- A digraph contains an Eulerian path if and only if it is connected and for the degrees of the vertices we have

$$d^+(v) = d^-(v) \text{ for all } v \neq v_1 \text{ or } v_2$$
$$d^+(v_1) = d^-(v_1) + 1$$
$$d^-(v_2) = d^+(v_2) + 1$$

Example 11.5   When we recall the Königsberg bridge problem, we see that the question is whether there exists an Eulerian circuit in the graph of Figure 11.2. The graph is connected. However, there are four vertices with odd degree. Therefore, the problem cannot be solved.

Example 11.6   A famous problem in graph theory is the *Chinese postman problem*. In its colloquial form it is about a postman who has to deliver the mail. In order to be more efficient, the question is whether he can traverse the street network of his town in such a way that starting at the post office he walks every street not more than once before he returns to the office. In graph theoretic terms, we are checking whether there is an Eulerian circuit in the graph of the street network.

11.4.2 Hamiltonian Tours

A graph is called Hamiltonian if it contains a Hamiltonian circuit. Unlike Eulerian graphs we do not have a simple way of determining whether a graph is Hamiltonian. All known algorithms to find a Hamiltonian tour in a graph are either inefficient or solving the problem only through approximations.

Example 11.7 A generalization of the Hamiltonian tour problem is the *traveling salesman problem*. The problem can be formulated as: Given a number of cities and the costs of traveling form one city to the other, what is the cheapest roundtrip that visits every city and then returns to the starting city? The cities are the vertices of a weighted graph. There are no efficient algorithms to solve the problem. Good approximations exist, however.

11.4.3 Shortest Path Problem

The shortest path problem can be formulated as follows. Given a weighted graph with $f: E \rightarrow \mathbb{R}$ assigning weights to the edges, and two vertices v_1 and v_2, find a path P from v_1 to v_2 such that for all edges $e \in P$ of the path $\Sigma_{e \in P} f(e)$ is minimal among all paths connecting v_1 and v_2.

Example 11.8 A well-known algorithm to solve the shortest path problem for a connected digraph with non-negative weights is Dijkstra's algorithm named after the Dutch computer scientist Edsger W. Dijkstra.

11.5 Applications in GIS

Graphs have played an important role in GIS right from the early beginnings. The reason is that in the early days of GIS the storage of map data (or cartographic data) was the focus of interest. Early data structures for the representation of spatial data (predominantly two-dimensional) are almost exclusively based on planar graphs.

One of the best-known examples is the GBF/DIME (Geographic Base File/ Dual Independent Map Encoding) file of the United States Census Bureau. This file structure was introduced to conduct the 1970 census. The United

States Geological Survey developed the DLG (Digital Line Graph) file format to store and transfer topographic base data.

In terms of data modeling planar graphs have been used extensively to represent two-dimensional spatial data. So-called *topological graphs* are the backbone of efficient representations. A topological graph is isomorphic to a planar graph embedded in $\mathbb{R}^2$. The vertices are usually called nodes, the edges are called arcs and the faces are called polygons. Such a topological graph is homeomorphic to a two-dimensional cell complex. A *network* consisting of nodes (vertices) and arcs (edges) can be considered a graph or a one-dimensional cell complex. Therefore, planar graphs or cell complexes can be used interchangeably as long as we do not exceed the two-dimensional space. For three-dimensional configurations we need to turn to topology. Besides the representation of spatial features, graphs play an important role in the representation and analysis of networks.

Definition 11.2 (Network). A *network* is a finite connected digraph in which one vertex x with $d^+(x) > 0$ is the *source* of the network, and one vertex y with $d^-(y) > 0$ is the *sink* of the network.

The network analysis functions of a GIS provide tools to find the shortest path from A to B, to perform allocation analysis, to trace a network path, as well as location-allocation analyses.

11.6   Exercises

Exercise 11.1   Draw K_5.

Exercise 11.2   Show that the spatial configuration of Exercise 9.1 is a planar graph using Euler's formula.

Chapter 12

Fuzzy Logic and GIS

Many phenomena show a degree of vagueness or uncertainty that cannot be properly expressed with crisp sets having sharp class boundaries. Spatial features often do not have clearly defined boundaries, and concepts like "steep", "close", or "suitable" can better be expressed with degrees of membership to a fuzzy set than with a binary yes/ no classification. This chapter introduces the basic principles of fuzzy logic, a mathematical theory that has found many applications in various domains. It can be applied whenever vague phenomena are involved.

12.1   Fuzziness

In human thinking and language, we often use uncertain or vague concepts. Our thinking and language are not binary, i.e., black and white, zero or one, yes or no. In real life, we add much more variation to our judgments and classifications. These vague or uncertain concepts are said to be fuzzy. We encounter fuzziness almost everywhere in our everyday lives.

12.1.1   Motivation

When we talk about tall people, the concept of "tall" will be depending on the context. In a society where the average height of a person is 160 cm, somebody will be considered to be tall differently from a population with an average height of 180 cm. In land cover analysis we are not able to draw crisp boundaries of, for instance, forest areas or grassland. Where does the grassland end and the forest start? The boundaries will be vague or fuzzy.

In real life applications, we might look for a suitable site to build a house. The criteria for the site that we are looking for could be formulated as follows. The site must

- have *moderate* slope;
- have *favorable* aspect;
- have *moderate* elevation;
- be *close to* a lake;
- be *not near* a major road;
- not be located in a restricted area.

All the conditions mentioned above (except the one for the restricted area) are vague, but correspond to the way we express these conditions in our languages and thinking. Using the conventional approach, the above-mentioned conditions would be translated into crisp classes, such as

- slope less than 10°;
- aspect between 135° and 225°, or the terrain is flat;
- elevation between 1,500 m and 2,000 m;
- within 1 km from a lake;
- not within 300 m from a major road.

If a location falls within the given criteria we would accept it, otherwise (even if it would be very close to the set threshold) it would be excluded from our analysis. If, however, we allow degrees of membership to our classes, we can accommodate also those locations that just miss a criterion by a few meters. They will just get a lower degree of membership, but will be included in the analysis. Usually, we assign a degree of membership to a class as a value between zero and one, where zero indicates no membership and one represents full membership. Any value in between can be a possible degree of membership.

12.1.2   Fuzziness versus Probability

Degrees of membership as values ranging between zero and one look very similar to probabilities, which are also given as a value between zero and one. We might be tempted to assume that fuzziness and probability are basically the same. There is, however, a subtle, yet important, difference.

Probability gives us an indication with which likelihood an event will occur. Whether it is going to happen, is not sure depending on the probability. *Fuzziness* is an indication to what degree something belongs to a class (or phenomenon). We know that the phenomenon exists. What we do not know, however, is its extent, i.e., to which degree members of a given universe belong to the class. In the following sections, we will establish the mathematical basis to deal with vague and fuzzy concepts.

12.2 Crisp Sets and Fuzzy Sets

In general set theory an element is either a member of a set or not. We can express this fact with the characteristic function for the elements of a given universe to belong to a certain subset of this universe. We call such a set a *crisp set*.

Definition 12.1 (Characteristic Function). Let A be a subset of a universe X. The *characteristic function* χ_A of A is defined as $\chi_A : X \rightarrow \{0, 1\}$ with

$$\chi_A(x) = \begin{cases} 1 & \text{iff } x \in A \\ 0 & \text{iff } x \notin A \end{cases}$$

In this way, we always can clearly indicate whether an element belongs to a set or not. If, however, we allow some degree of uncertainty as to whether an element belongs to a set, we can express the membership of an element to a set by its membership function.

Definition 12.2 (Fuzzy Set). A *fuzzy set* A of a universe X is defined by a *membership function* μ_A such that $\mu_A : X \rightarrow [0, 1]$, where $\mu_A(x)$ is the *membership value* of x in A. The universe X is always a crisp set.

If the universe is a finite set $X = \{x_1, x_2, \ldots, x_n\}$, then a fuzzy set A on X is expressed as $A = \mu_A(x_1)/x_1 + \mu_A(x_2)/x_2 + \cdots + \mu_A(x_n)/x_n = \sum_{i=1}^{n} \mu_A(x_i)/x_i$. The term $\mu_A(x_i)/x_i$ indicates the membership value to fuzzy set A for x_i. The symbol "/" is called *separator*, "Σ" and "+" function as *aggregation* and *connection* of terms.[1]

---

[1]Note that the symbols "Σ", "+", and "$\int$" are not interpreted in their usual meanings as sum, addition, and integral.

If the universe is an infinite set $X = \{x_1, x_2, ...\}$, then a fuzzy set A on X is expressed as $A = \int_X \mu_A(x)/x$. The symbols "$\int$" and "$/$" function as aggregation and separator.

The *empty fuzzy set* $\emptyset$ is defined as $\forall x \in X, \mu_\emptyset(x) = 0$.

For every element of the universe X we trivially have $\forall x \in X, \mu_X(x) = 1$, i.e., the universe is always crisp.

A membership function assigns to every element of the universe a degree of membership (or membership value) to a fuzzy set. This membership value must be between zero (no membership) and one (definite membership). All other values between zero and one indicate to which degree an element belongs to the fuzzy set. It is important to note that the degree of membership of 1 does not need to be obtained for members of a fuzzy set.

Example 12.1 Let us take three persons A, B, and C and their respective heights as $185\,\text{cm}(A)$, $165\,\text{cm}(B)$ and $186\,\text{cm}(C)$. We want to assign the different persons to classes for short, average, and tall people, respectively.

If we take a crisp classification and set the class boundaries to $(-, 165]$ for short, $(165, 185]$ for average, and $(185, -)$ for tall, we see that A falls into the average class, B falls into the short class, and C falls into the tall class. We also see that A is nearly as tall as C, and yet they fall into different classes. The characteristic functions of the three classes are displayed in Table 12.1.

Table 12.1 Characteristic functions for the height classes

Person	Short	Average	Tall
A	0	1	0
B	1	0	0
C	0	0	1

When we choose a fuzzy set approach, we need to define three membership functions for the three classes, respectively (Figure 12.1).

For *short* we select a linear membership function that produces a membership value of one for persons shorter than 150 cm and decreases until it reaches zero at 180 cm.

The membership function for the *average* class produces values equal to zero for persons shorter than 150 cm. It then increases until it reaches one at 175 cm. From there it decreases until it reaches zero at 200 cm.

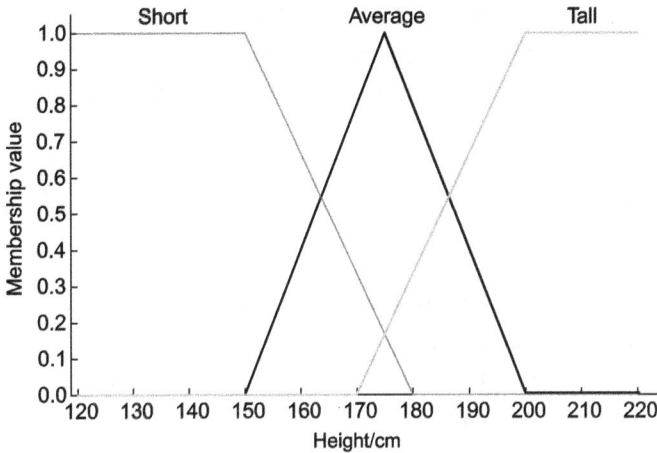

Figure 12.1 Membership functions for "short", "average", and "tall"

The membership function for the *tall* class is zero up to 170 cm. From there it increases until it reaches one at 200 cm. The membership values for the three persons to the three classes are given in Table 12.2.

Table 12.2 Membership values for the height classes

	Short	Average	Tall
A	0.00	0.60	0.50
B	0.50	0.60	0.00
C	0.00	0.56	0.53

Using the fuzzy set approach, we can much better express the fact that *A* and *C* are nearly the same height and that both have a higher degree of membership to the average class than to the short or tall class, respectively.

12.3 Membership Functions

The selection of a suitable membership function for a fuzzy set is one of the most important activities in fuzzy logic. It is the responsibility of the user to select a function that is a best representation for the fuzzy concept to be modeled. The following criteria are valid for all membership functions:

- The membership function must be a real valued function whose values are between 0 and 1.

- The membership values should be 1 at the center of the set, i.e., for those members that definitely belong to the set.
- The membership function should fall off in an appropriate way from the center to the boundary.
- The points with membership value 0.5 (crossover point) should be at the boundary of the crisp set, i.e., if we would apply a crisp classification, the class boundary should be represented by the crossover points.

We know two types of membership functions: (i) linear membership functions and (ii) sinusoidal membership functions. Figure 12.2 shows the linear membership function. This function has four parameters that determine the shape of the function. By choosing proper values for a, b, c, and d, we can create S-shaped, trapezoidal, triangular, and L-shaped membership functions.

If a nonlinear shape of the membership function is more appropriate for our purpose we should choose a sinusoidal membership function (Figure 12.3). As with linear membership functions we can achieve S-shaped,

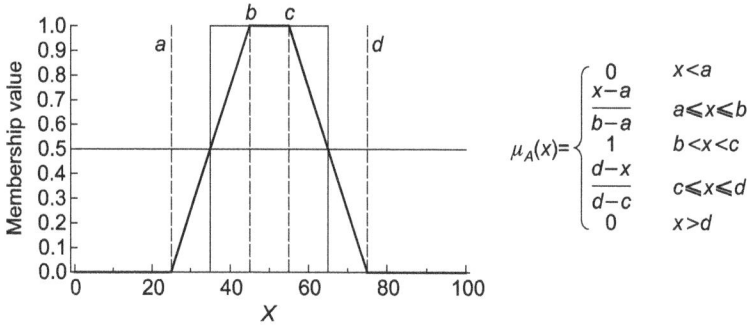

$$\mu_A(x)=\begin{cases} 0 & x<a \\ \dfrac{x-a}{b-a} & a\leqslant x\leqslant b \\ 1 & b<x<c \\ \dfrac{d-x}{d-c} & c\leqslant x\leqslant d \\ 0 & x>d \end{cases}$$

Figure 12.2   Linear membership function

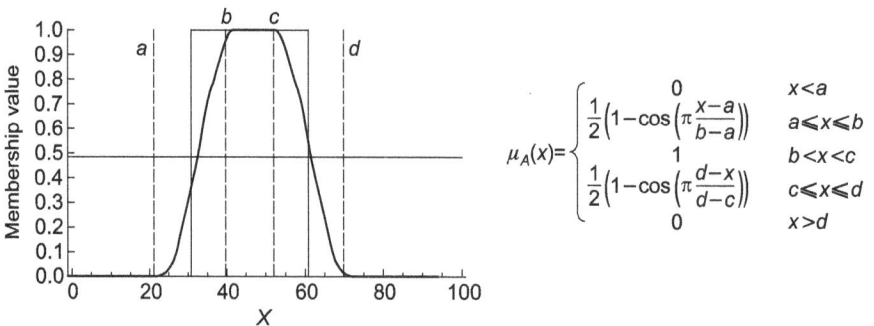

$$\mu_A(x)=\begin{cases} 0 & x<a \\ \dfrac{1}{2}\left(1-\cos\left(\pi\dfrac{x-a}{b-a}\right)\right) & a\leqslant x\leqslant b \\ 1 & b<x<c \\ \dfrac{1}{2}\left(1-\cos\left(\pi\dfrac{d-x}{d-c}\right)\right) & c\leqslant x\leqslant d \\ 0 & x>d \end{cases}$$

Figure 12.3   Sinusoidal membership function

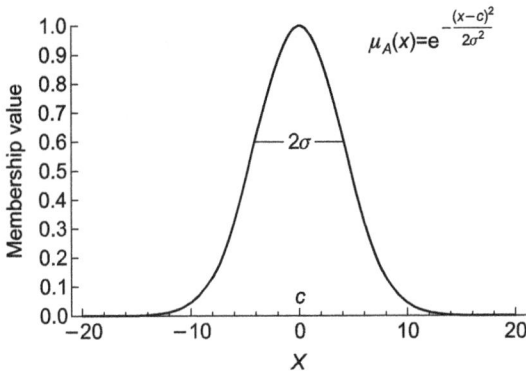

Figure 12.4   Gaussian membership function

bell-shaped, and L-shaped membership functions by proper selection of the four parameters.

A special case of the bell-shaped membership functions is the Gaussian function (Figure 12.4) derived from the probability density function of the normal distribution with two parameters c (mean) and σ (standard deviation). Although this membership function is derived from probability theory, it is used here as a membership function for a fuzzy set.

Example 12.2 The membership functions in Example 12.1 are linear functions with the following parameters:

$$\mu_{short}(x) = \begin{cases} 1 & x \leqslant 150 \\ \dfrac{180 - x}{30} & 150 < x \leqslant 180 \\ 0 & x > 180 \end{cases}$$

$$\mu_{average}(x) = \begin{cases} 0 & x \leqslant 150 \\ \dfrac{x - 150}{25} & 150 < x \leqslant 175 \\ \dfrac{200 - x}{25} & 175 \leqslant x \leqslant 200 \\ 0 & x > 200 \end{cases}$$

$$\mu_{tall}(x) = \begin{cases} 0 & x \leqslant 170 \\ \dfrac{x - 170}{30} & 170 < x \leqslant 200 \\ 1 & x > 200 \end{cases}$$

12.4   Operations on Fuzzy Sets

Operations on fuzzy sets are defined in a similar way as for crisp sets. However, not all rules for crisp set operations are also valid for fuzzy sets. Like for crisp sets we have subset, union, intersection, and complement. In addition, there are alternate operations for union and intersection of fuzzy sets.

Definition 12.3 (Support). All elements of the universe X that have membership values greater than zero for a fuzzy set A are called the *support* of A, or $\text{supp}(A) = \{x \in X | \mu_A(x) > 0\}$.

Example 12.3   The support of the fuzzy set for short people (Example 12.1) is those persons who are shorter than 180 cm.

Definition 12.4 (Height). The *height* of a fuzzy set A is the largest membership value in A, written as $\text{hgt}(A)$. If $\text{hgt}(A) = 1$ then the set is called *normal*.

Example 12.4   The height of the fuzzy sets Short, Average, and Tall is 1. They are all normal fuzzy sets.

We can always normalize a fuzzy set by dividing all its membership values by the height of the set.

Definition 12.5 (Equality).   Two fuzzy sets A and B are *equal* (written as $A = B$) if for all members of the universe X their membership values are equal, i.e., $\forall x \in X, \mu_A(x) = \mu_B(x)$.

Subsets in fuzzy sets are defined by fuzzy set inclusion.

Definition 12.6 (Inclusion).   A fuzzy set A is *included* in a fuzzy set B (written as $A \subseteq B$) if for every element of the universe the membership values for A are less than or equal to those for B, i.e., $\forall x \in X, \mu_A(x) \leqslant \mu_B(x)$.

When we look at the graph of the membership functions a fuzzy set A will be included in fuzzy set B when the graph of A is completely covered by the graph of B (Figure 12.5).

For the union of two fuzzy sets we have more than one operator. The most common ones are presented here.

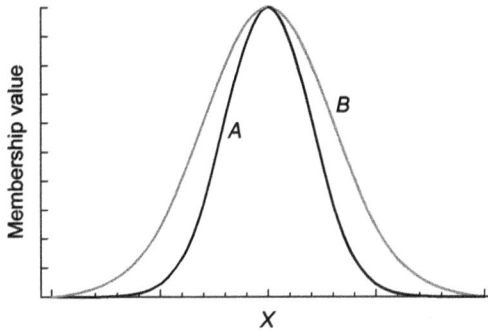

Figure 12.5   Set inclusion

Definition 12.7 (Union). The *union* of two fuzzy sets A and B can be computed for all elements of the universe X by one of the three operators:

(1) $\mu_{A \cup B}(x) = \max(\mu_A(x), \mu_B(x))$.

(2) $\mu_{A \cup B}(x) = \mu_A(x) + \mu_B(x) - \mu_A(x) \cdot \mu_B(x)$.

(3) $\mu_{A \cup B}(x) = \min(1, \mu_A(x) + \mu_B(x))$.

The max-operator is a *non-interactive* operator in the sense that the membership values of both sets do not interact with each other. In fact, one set could be completely ignored in a union operation when it is included in the other. The two other operators are called interactive, because the membership value of the union is determined by the membership values of both sets.

Example 12.5   Figure 12.6 illustrates the union operators for the fuzzy sets Short and Average from Example 12.1.

Definition 12.8 (Intersection). The *intersection* of two fuzzy sets A and B can be computed for all elements of the universe X by one of the three operators:

(1) $\mu_{A \cap B}(x) = \min(\mu_A(x), \mu_B(x))$.

(2) $\mu_{A \cap B}(x) = \mu_A(x) \cdot \mu_B(x)$.

(3) $\mu_{A \cap B}(x) = \max(0, \mu_A(x) + \mu_B(x) - 1)$.

The min-operator is a non-interactive operator, and the two others are interactive operators as explained above.

Example 12.6   Figure 12.7 illustrates the intersection operators for the fuzzy sets Short and Average from Example 12.1.

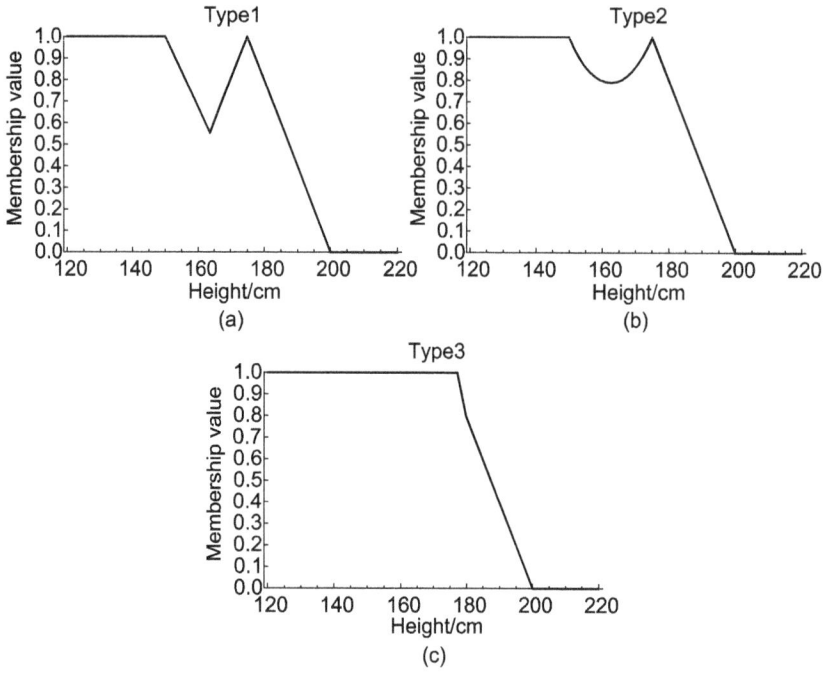

Figure 12.6   Fuzzy set union operators

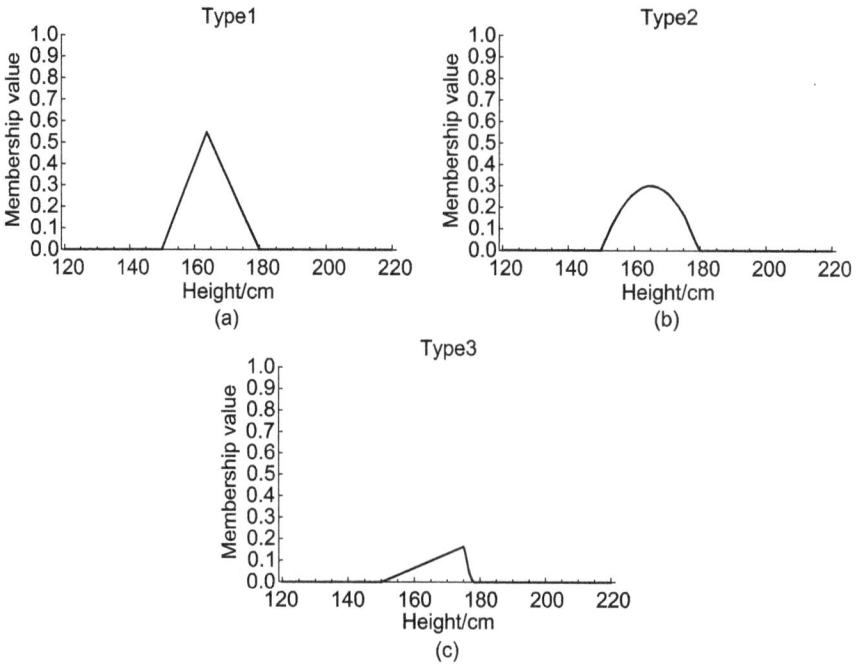

Figure 12.7   Fuzzy set intersection operators

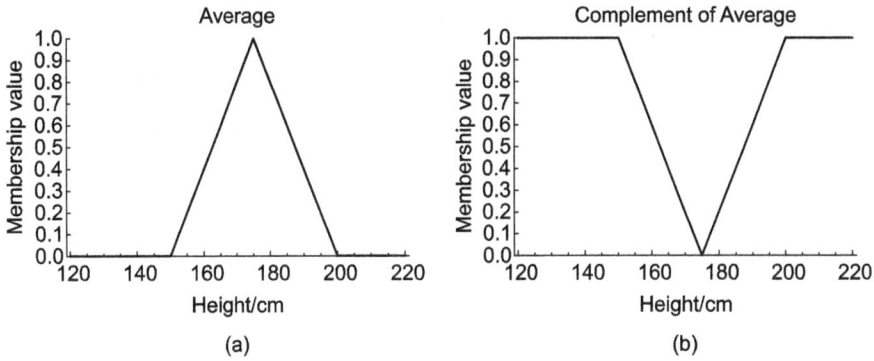

Figure 12.8 Fuzzy set and its complement

Definition 12.9 (Complement). The *complement* of a fuzzy set A in the universe X is defined as $\forall x \in X$, $\mu_{\overline{A}}(x) = 1 - \mu_A(x)$.

Example 12.7 Figure 12.8 shows the fuzzy set Average from Example 12.1 and its complement.

Many rules for set operations are valid for both crisp and fuzzy sets. Table 12.3 shows the rules that are valid for both.

Table 12.3 Rules for set operations valid for crisp and fuzzy sets

(1) $A \cup A = A$	
(2) $A \cap A = A$	Idempotent law
(3) $(A \cup B) \cup C = A \cup (B \cup C)$	
(4) $(A \cap B) \cap C = A \cap (B \cap C)$	Associativity
(5) $A \cup B = B \cup A$	
(6) $A \cap B = B \cap A$	Commutativity
(7) $A \cup (B \cap C) = (A \cup B) \cap (A \cup C)$	
(8) $A \cap (B \cup C) = (A \cap B) \cup (A \cap C)$	Distributivity
(9) $\overline{A \cup B} = \overline{A} \cap \overline{B}$	
(10) $\overline{A \cap B} = \overline{A} \cup \overline{B}$	De Morgan's Laws
(11) $\overline{\overline{A}} = A$	Double complement

Table 12.4 shows those rules that in general are valid for crisp sets but not for fuzzy sets.

Table 12.4 Rules valid only for crisp sets

(1) $A \cup \bar{A} = X$	Law of the excluded middle
(2) $A \cap \bar{A} = \emptyset$	Law of contradiction

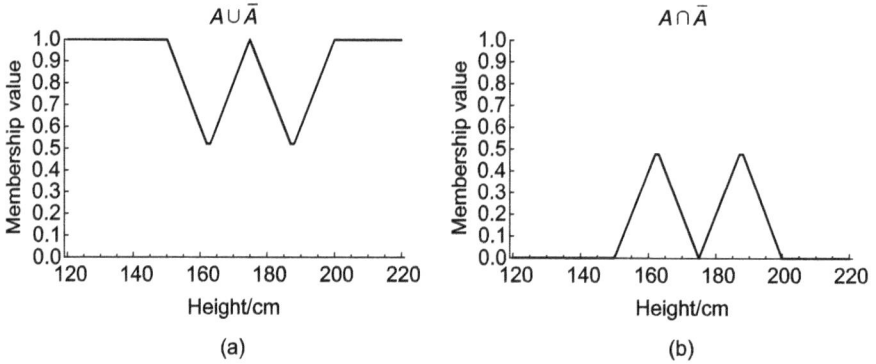

Figure 12.9 Law of the excluded middle and law of contraction for fuzzy set Average.

Figure 12.9 illustrates that the law of the excluded middle and the law of contradiction does not generally hold for fuzzy sets.

12.5 α-Cuts, α-Level Set

If we wish to know all those elements of the universe that belong to a fuzzy set and have at least a certain degree of membership, we can use α-level sets.

Definition 12.10 (α-Cut). A *weak α-cut* (or *α-level set*) A_α with $0 < \alpha \leqslant 1$ is the set of all elements of the universe such that $\mu_A(x) \geqslant \alpha$, i.e., $A_\alpha = \{x \in X | \mu_A(x) \geqslant \alpha\}$. A *strong α-cut* $A_{\bar{\alpha}}$ is defined as $A_{\bar{\alpha}} = \{x \in X | \mu_A(x) > \alpha\}$.

Example 12.8   The 0.8-cut of the fuzzy set Tall contains all those persons who are 194 cm or taller.

With α-level sets we can identify those members of the universe that typically belong to a fuzzy set.

12.6   Linguistic Variables and Hedges

In mathematics variables usually assume numbers as values. A *linguistic variable* is a variable that assumes linguistic values which are words (*linguistic terms*). If, for example, we have the linguistic variable "height", the linguistic values for height could be "short", "average", and "tall". These linguistic values possess a certain degree of uncertainty or vagueness that can be expressed by a membership function of a fuzzy set. Often, we modify a linguistic term by adding words like "very", "somewhat", "slightly", or "more or less" and arrive at expressions such as "very tall", "not short", or "somewhat average".

Such modifiers are called *hedges*. They can be expressed with operators applied to the fuzzy sets representing linguistic terms (see Table 12.5).

Table 12.6 shows the models used to represent hedges for linguistic terms.

Table 12.5   Operators for hedges

Operator	Expression	
Normalization	$\mu_{\text{norm}(A)}(x) = \dfrac{\mu_A(x)}{\text{hgt}(A)}$	
Concentration	$\mu_{\text{con}(A)}(x) = \mu_A^2(x)$	
Dilation	$\mu_{\text{dil}(A)}(x) = \sqrt{\mu_A(x)}$	
Negation	$\mu_{\text{not}(A)}(x) = \mu_{\overline{A}}(x) = 1 - \mu_A(x)$	
Contrast intensification	$\mu_{\text{int}(A)}(x) = \begin{cases} 2\mu_A^2(x) & \\ 1 - 2\left(1 - \mu_A(x)\right)^2 & \end{cases}$	if $\mu_A(x) \in [0, 0.5]$ otherwise

Table 12.6   Hedges and their models

Hedge	Operator
A little A	$[\mu_A(x)]^{1.3}$
Slightly A	$[\mu_A(x)]^{1.7}$
Very A	$[\mu_A(x)]^2$
Extremely A	$[\mu_A(x)]^3$
Very very A	$[\mu_A(x)]^4$
More or less A (somewhat A)	$\sqrt{\mu_A(x)}$
Slightly A	Contrast intensification $(\mu_A(x))$
Not A	$1 - \mu_A(x)$
Comfortably A (reasonably A)	int(norm(slightly $A \cap$ not(very A)))

Example 12.9 Figure 12.10 shows the membership functions for Tall, Very Tall, and Very Very Tall.

Example 12.10 Figure 12.11 shows the membership functions for Tall and Not Very Tall.

Example 12.11 Figure 12.12 shows the membership functions for Tall and Reasonably Tall.

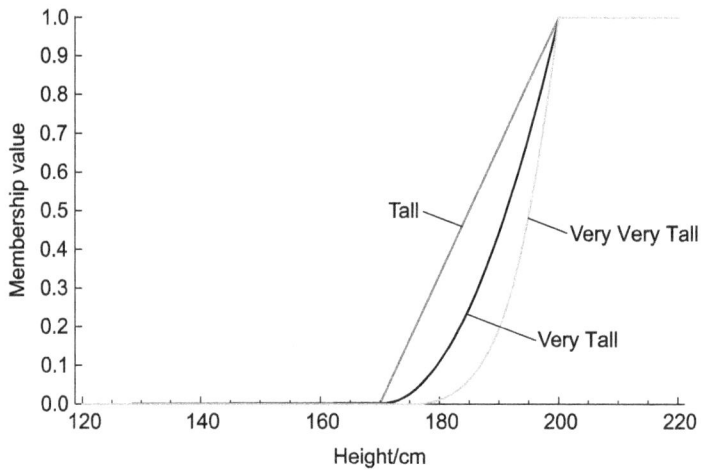

Figure 12.10 Membership functions for Tall, Very Tall, and Very Very Tall

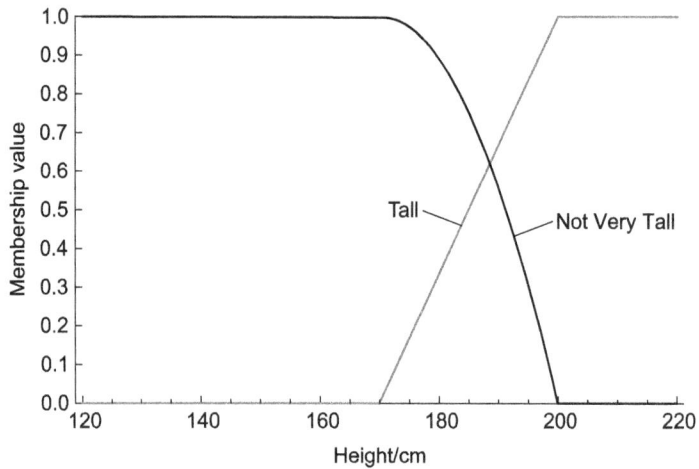

Figure 12.11 Membership functions for Tall and Not Very Tall

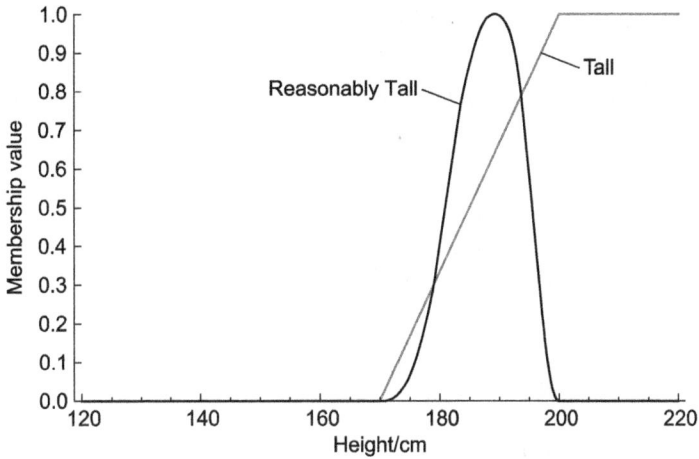

Figure 12.12   Membership functions for Tall and Reasonably Tall

12.7   Fuzzy Inference

In binary logic we have only two possible values for a logical variable, true or false, 1 or 0. As we have seen in this chapter, many phenomena can be better represented by fuzzy sets than by crisp sets. Fuzzy sets can also be applied to reasoning when vague concepts are involved.

In binary logic reasoning is based on either deduction (*modus ponens*) or induction (*modus tollens*). In fuzzy reasoning we use a *generalized modus ponens* which reads as

Premise 1:   If x is A then y is B
Premise 2: x is A'
Conclusion: y is B'

Here, A, B, A', and B' are fuzzy sets where A' and B' are not exactly the same as A and B.

Example 12.12   Consider the generalized modus ponens for temperature control:

Premise 1:   If the temperature is *low* then set the heater to *high*
Premise 2:   Temperature is *very low*
Conclusion: Set the heater to *very high*

With logic inference we normally have more than one rule. In fact, the number of rules can be rather large. We know several methods for fuzzy reasoning.

12.7.1 Mamdani's Direct Method

Here, we discuss the method known as Mamdani's direct method. It is based on a generalized modus ponens of the form

$$p \Rightarrow q : \begin{cases} \text{If } x \text{ is } A_1 \text{ and } y \text{ is } B_1 \text{ then } z \text{ is } C_1 \\ \text{If } x \text{ is } A_2 \text{ and } y \text{ is } B_2 \text{ then } z \text{ is } C_2 \\ \quad \vdots \\ \text{If } x \text{ is } A_n \text{ and } y \text{ is } B_n \text{ then } z \text{ is } C_n \end{cases}$$

$$\frac{p_1 : \qquad\qquad x \text{ is } A', y \text{ is } B'}{q_1 : \qquad\qquad z \text{ is } C'}$$

Premise 1 becomes a set of rules as illustrated in Figure 12.13. A, B, and C are fuzzy sets, x and y are the *premise variables*, and z is the *consequence variable*.[2]

The reasoning process is then straightforward according to the following procedure. Let x_0 and y_0 be the input for the premise variables.

(1) Apply the input values to the premise variables for every rule and compute the minimum of $\mu_{A_i}(x_0)$ and $\mu_{B_i}(y_0)$:

Rule 1: $\quad m_1 = \min(\mu_{A_1}(x_0), \mu_{B_1}(y_0))$
Rule 2: $\quad m_2 = \min(\mu_{A_2}(x_0), \mu_{B_2}(y_0))$
$\quad \vdots \qquad\qquad \vdots$
Rule n: $\quad m_n = \min(\mu_{A_n}(x_0), \mu_{B_n}(y_0))$

(2) Cut the membership function of the consequence $\mu_{C_i}(z)$ at m_i:

Conclusion of Rule 1: $\quad \mu_{C_1'} = \min(m_1, \mu_{C_1}(z)) \forall z \in C_1$
Conclusion of Rule 2: $\quad \mu_{C_2'} = \min(m_2, \mu_{C_2}(z)) \forall z \in C_2$
$\quad \vdots \qquad\qquad\qquad\qquad \vdots$
Conclusion of Rule n: $\quad \mu_{C_n'} = \min(m_n, \mu_{C_n}(z)) \forall z \in C_n$

(3) Compute the final conclusion by determining the union of all individual conclusions from Step 2:

$$\mu_C(z) = \max(\mu_{C_1'}(z), \mu_{C_2'}(z), \dots, \mu_{C_n'}(z))$$

---

[2]There can be more than two premise variables to express complex rules. The procedure can be extended to this case without any problems.

If *x* is *A* and *y* is *B* then *z* is *C*

Premise   Consequence

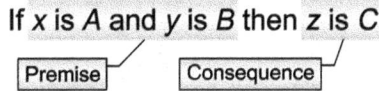

Figure 12.13   Inference rule in Mamdani's direct method

The result of the final conclusion is a fuzzy set. For practical reasons, we need a definite value for the consequence variable. The process to determine this value is called *defuzzification*. There are several methods to defuzzify a given fuzzy set. One of the most common methods is the *center of gravity* (or center of area).

For a discrete fuzzy set the center of area is computed as

$$z_0 = \frac{\sum \mu_C(z) \cdot z}{\sum \mu_C(z)}$$

For a continuous fuzzy set this becomes

$$z_0 = \frac{\int \mu_C(z) \cdot z dz}{\int \mu_C(z) dz}$$

Example 12.13   Given the speed of a car and the distance to a car in front of it, we would like to determine whether we should break, maintain the speed, or accelerate. Assume the following set of rules for the given situation:

Rule 1   If the distance between the cars is short and the speed is low then maintain speed.

Rule 2   If the distance between the cars is short and the speed is high then reduce speed.

Rule 3   If the distance between the cars is long and the speed is low then increase speed.

Rule 4   If the distance between the cars is long and the speed is high then maintain speed.

Distance, speed, and acceleration are linguistic variables with the values "short", "long", "high", "low", and "reduce", "maintain", and "increase", respectively. They can be modeled as fuzzy sets (Figure 12.14).

With a given distance $x_0 = 15\,\text{m}$ and a speed of $y_0 = 60\,\text{km} \cdot \text{h}^{-1}$ we perform Step 1. The results are shown in Table 12.7.

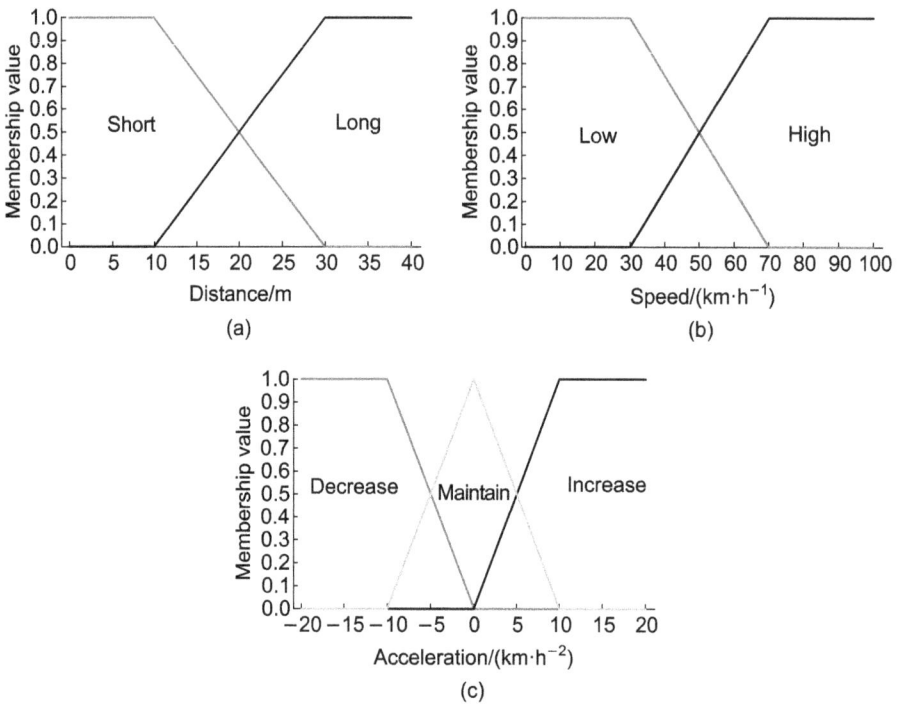

Figure 12.14   Fuzzy sets of the rules

Table 12.7   Fuzzy inference Step 1

Rule	Short	Long	Low	High	Min
1	0.75		0.25		0.25
2	0.75			0.75	0.75
3		0.25	0.25		0.25
4		0.25		0.75	0.25

Now, we must cut the membership function for the conclusion variable at the minimum values from Step 1. The result is illustrated in Figure 12.15.

Finally, we must combine the individual membership functions from Step 2 to the final result and defuzzify it. The union of the four membership functions is displayed in Figure 12.16. The final value after defuzzification is −5.46 and is indicated by the dot. The conclusion of this fuzzy inference is that when the distance between the cars is 15 m and the speed is 60 km·h^{-1}, then we have to break gently to reduce the speed.

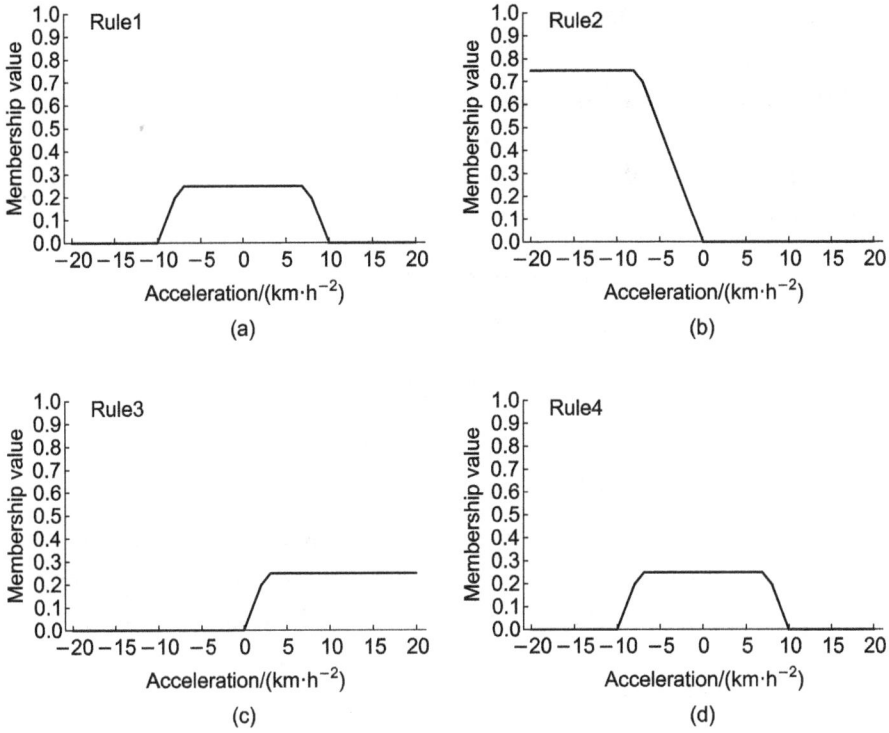

Figure 12.15   Fuzzy inference Step 2

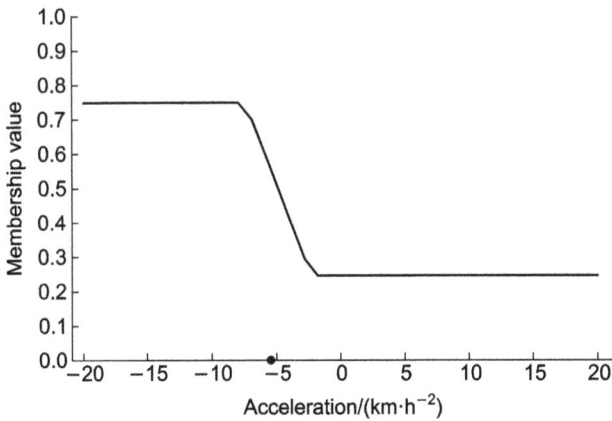

Figure 12.16   Fuzzy inference final result

12.7.2 Simplified Method

Often, the defuzzification process is too time-consuming and complicated. An alternative approach is the simplified method where the conclusion is a real value c instead of a fuzzy set. It is based on a generalized *modus ponens* of the form:

$$p \Rightarrow q : \begin{cases} \text{If } x \text{ is } A_1 \text{ and } y \text{ is } B_1 \text{ then } z \text{ is } c_1 \\ \text{If } x \text{ is } A_2 \text{ and } y \text{ is } B_2 \text{ then } z \text{ is } c_2 \\ \quad \vdots \\ \text{If } x \text{ is } A_n \text{ and } y \text{ is } B_n \text{ then } z \text{ is } c_n \end{cases}$$

$$\frac{p_1 : \qquad\qquad x \text{ is } A', y \text{ is } B'}{q_1 : \qquad\qquad z \text{ is } c'}$$

Premise 1 becomes a set of rules as illustrated in Figure 12.17. The premise variables are fuzzy sets; the conclusion is a real number (fuzzy singleton).

The reasoning process is then straightforward in analogy to the previous method with the difference that the result is not a fuzzy set that needs to be defuzzified but can be deduced by computing the final result directly after Step 2 in the algorithm.

The algorithm works as outlined in the following procedure. Let x_0 and y_0 be the input for the premise variables.

(1) Apply the input values to the premise variables for every rule and compute the minimum of $\mu_{A_i}(x_0)$ and $\mu_{B_i}(y_0)$:

Rule$_1$: $m_1 = \min(\mu_{A_1}(x_0), \mu_{B_1}(y_0))$
Rule$_2$: $m_2 = \min(\mu_{A_2}(x_0), \mu_{B_2}(y_0))$
  $\vdots$ $\vdots$
Rule$_n$: $m_n = \min(\mu_{A_n}(x_0), \mu_{B_n}(y_0))$

(2) Compute the conclusion value per rule as:

Conclusion of Rule$_1$: $c'_1 = m_1 \cdot c_1$
Conclusion of Rule$_2$: $c'_2 = m_2 \cdot c_2$
  $\vdots$ $\vdots$
Conclusion of Rule$_n$: $c'_n = m_n \cdot c_n$

(3) Compute the final conclusion as:

$$c' = \frac{\sum_{i=1}^{n} c'_i}{\sum_{i=1}^{n} m_i}$$

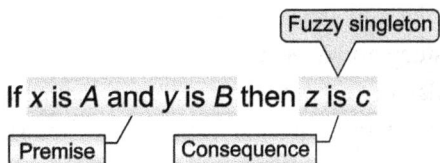

Figure 12.17   Inference rule in simplified Method

Example 12.14   Given the slope and aspect maps of a region and the following set of rules, we can conduct a risk analysis based on degrees of risk ranging from 1 (low risk) to 4 (very high risk). The fuzzy sets for flat and steep slope are displayed in Figures 12.18 and 12.19.

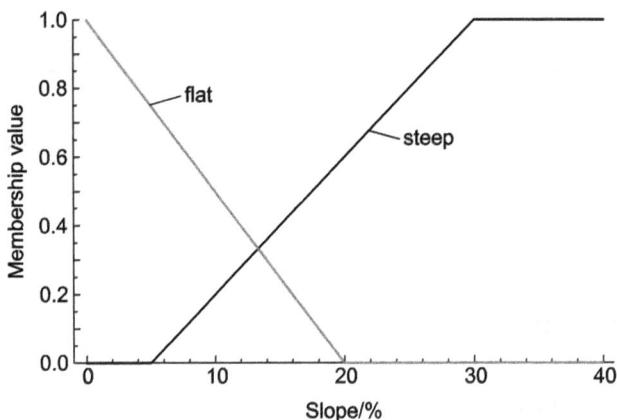

Figure 12.18   Membership functions for flat and steep slopes

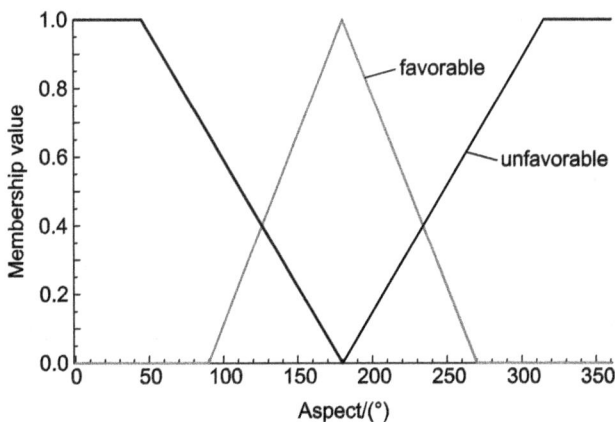

Figure 12.19   Membership functions for favorable and unfavorable aspects.

Rule 1   If slopes is flat and aspect is favorable then risk is 1.
Rule 2   If slope is steep and aspect is favorable then risk is 2.
Rule 3   If slope is flat and aspect is unfavorable then risk is 1.
Rule 4   If slope is steep and aspect is unfavorable then risk is 4.

For a slope of 10 percent and an aspect of 180° we have the following results (Table 12.8):

Table 12.8   Reasoning results of the simplified method

	Slope (s)	Aspect (a)	Min(s,a)	Conclusions
Rule 1	0.5	1	0.5	0.5
Rule 2	0.2	1	0.2	0.4
Rule 3	0.5	0	0	0
Rule 4	0.2	0	0	0

For the final result we get $c' = \dfrac{0.5 + 0.4 + 0 + 0}{0.5 + 0.2 + 0 + 0} = 1.29$, which means a low risk.

12.8   Applications in GIS

Many spatial phenomena are inherently fuzzy or vague or possess indeterminate boundaries. Fuzzy logic has been applied for many areas in GIS such as fuzzy spatial analysis, fuzzy reasoning, and the representation of fuzzy boundaries. The following example illustrates how a fuzzy set can be computed from a given grid data set.

12.8.1   Objective

The objective of this analysis is to determine high elevation. In the following illustrations we used a digital elevation model of the area covered by a USGS 1:24,000 topographic map sheet of Boulder, Colorado. Any elevation data set can be used.

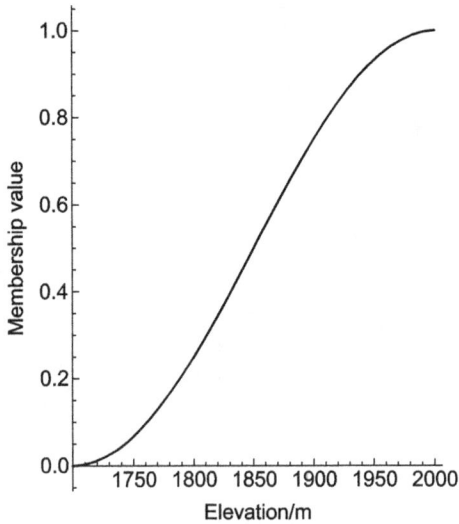

Figure 12.20   Membership function for "high elevation"

12.8.2   Fuzzy Concepts

Elevation is considered high when it is above 1,700 m. We represent the features meeting the criterion as a fuzzy set with a sinusoidal membership function (Figure 12.20) defined as

$$\mu_{\text{high elevation}}(x) = \begin{cases} 0 & x \leqslant 1700 \\ \dfrac{1}{2}\left(1 - \cos\left(\pi\dfrac{x - 1700}{300}\right)\right) & 1700 < x \leqslant 2000 \\ 1 & x > 2000 \end{cases}$$

12.8.3   Software Approach

The following examples are carried out with a digital elevation data set that is imported into ArcGIS. There are several ways to solve the problem: we can use ArcMap or ArcGIS Pro Spatial Analyst raster calculator tool or the Fuzzy Membership and Fuzzy Overlay tools. We can even create our own fuzzy logic tool using Python scripts. The elevation grid involved is called "elevation". The fuzzy set will be a grid "felevation" whose values are between zero and one.

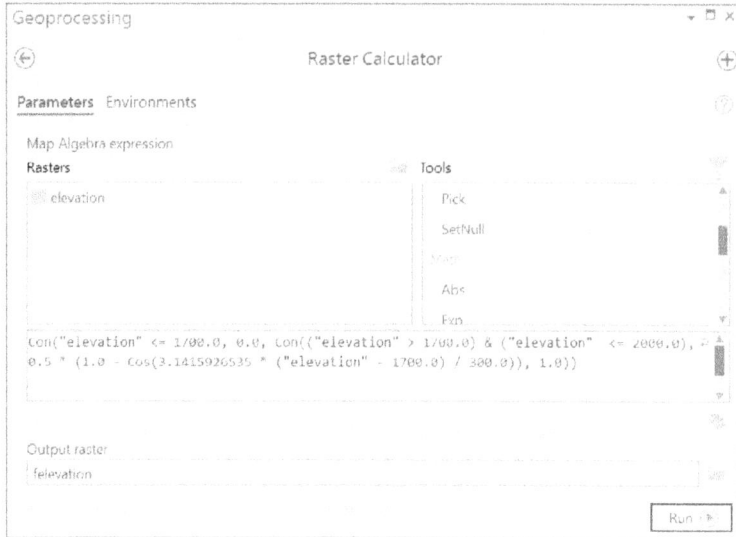

Figure 12.21   ArcGIS Pro Raster Calculator command for "high elevation"

12.8.3.1   *ArcGIS Pro Spatial Analyst Raster Calculator*

To solve the problem, we use the raster calculator of the Spatial Analyst. Figure 12.21 shows the ArcGIS Pro Raster Calculator command to produce the required fuzzy set.

12.8.3.2   *ArcGIS Pro Fuzzy Toolset*

We can also use the Fuzzy Membership tool of the Overlay Toolset in the Spatial Analyst. Here, however, we can only select from a given set of membership function types.

12.8.4   Result

Figure 12.22 shows the result of the analysis with a fuzzy logic approach and a crisp approach.

Fuzzy set   Crisp set

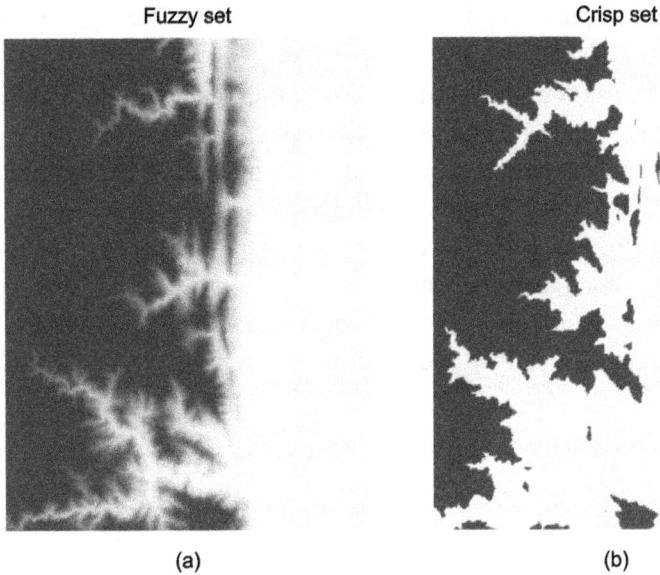

(a)   (b)

Figure 12.22 Analysis with a fuzzy logic approach (a) and a crisp approach (b)

12.9   Exercises

Exercise 12.1 Determine a linear membership function for "moderate elevation" when the ideal elevation is between 400 m and 600 m.

Exercise 12.2 Determine a Gaussian membership function for the aspect "south".

Exercise 12.3 Use a topographic data set with elevation, water bodies, and roads and determine a suitable site with the following characteristics:
- moderate slope
- favorable aspect
- moderate elevation
- near a lake or reservoir
- not very close to a major road

Choose suitable membership functions for the fuzzy terms.

Exercise 12.4 Design a simple fuzzy reasoning system for avalanche risk in a mountainous area. The variables involved are slope, aspect, and snow cover change. For simplicity we do not consider surface cover. The snow cover change must be simulated. The rules are given as:

Rule 1   If the slope is very steep and the aspect is unfavorable and the snow cover change is big then the risk is very high.

Rule 2   If the slope is moderate and the aspect is unfavorable and the snow cover change is big then the risk is moderate.

Rule 3   If the slope is steep and the aspect is unfavorable and the snow cover change is small then the risk is low.

Rule 4   If the slope is not steep and the aspect is unfavorable and the snow cover change is big then the risk is moderate.

Chapter 13

Probability Theory

Under a standard atmospheric pressure, pure water must be brought to boil when heated to 100°C. This phenomenon that a certain result must occur under certain conditions is called a decisive phenomenon. In contrast with the decisive phenomenon, tossing a coin may result in heads or tails. The phenomenon that when the basic conditions are not changed before each test or observation, it is not certain which kind of result will appear, and it shows chance, is called a random phenomenon.

Probability theory is a branch of mathematics that studies the quantitative law of random phenomena. Probability theory uses probability to measure the likelihood of the occurrence of an event. Although the occurrence of an event in a random experiment is accidental, many repeated random experiments under the same conditions often show obvious quantitative laws.

13.1 Random Event and Probability

13.1.1 Random Experiment and Sample Space

Definition 13.1 (Random Experiment). Probability theory studies statistical laws by repeated observations of a random phenomenon, or carrying out repeated experiments under the same conditions. Such an observation or an experiment is called a *random experiment*.

Random experiments have the following characteristics: (i) the experiment can be repeated under the same conditions; (ii) the result cannot be determined before the experiment; (iii) all the possible results of the experiment are explicit; (iv) for each experiment, when carried out, the result is explicit.

Typical random experiments are throwing a dice, tossing a coin, drawing a card, and playing roulette, etc.

Definition 13.2 (Sample Space). A *sample space*, usually denoted by Ω, is the set of all possible results of a random experiment.

Definition 13.3 (Sample). A *sample* is a result of a random experiment. It is an element of a sample space.

Definition 13.4 (Random Event). A subset of a sample space of a random experiment, is called a *random event*, or *event* for short. As two special subsets, the whole sample space is called a certain event, and a null set called a null event, usually denoted by Φ.

Since a random event is a set, random events follow set operators such as intersection, union, difference and complement, etc. The intersection of two events A and B, denoted by $A \cap B$, or AB, or $A * B$, is the event that both A and B occur. The union (sum) of two events A and B, denoted by $A \cup B$, or $A + B$, is the event that A or B occurs. The difference of an event A by another event B, denoted by $A - B$, or $A \backslash B$, is the event that A occurs and B does not occur. The complement of an event A, denoted by $\overline{A}$, is the event that A does not occur, i.e., $\Omega \backslash A$.

Example 13.1 By tossing a coin and observing the cases of head and tail, with the two basic results denoted by w_1 and w_2, the sample space of this random experiment is $\Omega = \{w_1, w_2\}$.

Example 13.2 By throwing a dice and observing the faces, the possible faces are 1, 2, 3, 4, 5, or 6. So, the sample space of this random experiment is $\Omega = \{1, 2, 3, 4, 5, 6\}$. Let $A = \{1, 2, 3\}$, then A is a random event of this sample space. Let $B = \{3, 4, 5, 6\}$, then B is another random event of this sample space. Now we have $AB = \{3\}$, $A + B = \Omega$, $A \backslash B = \{1, 2\}$, $\overline{A} = \{4, 5, 6\}$.

13.1.2 Probability

Definition 13.5 (Probability). Suppose Ω is the sample space of a random experiment. If for any event A, there is one and only one real number $P(A)$ corresponding to A, and the following axioms hold:

(1) Non-negative: $P(A) \geqslant 0$;

(2) Normality: $P(\Omega) = 1$;

(3) Conformable additivity: For any mutually pairwise exclusive events A_1,

$A_2, \ldots, A_n, \ldots$, there is, $P(\bigcup_{n=1}^{\infty} A_n) = \sum_{n=1}^{\infty} P(A_n)$;

then $P(A)$ is called the probability of the event A.

13.2 Conditional Probability and Independence

13.2.1 Conditional Probability

Definition 13.6 (Conditional Probability). The possibility of the occurrence of event A is affected by the occurrence of another related event B. After the event B occurs, the possibility of the occurrence of the event A is called the *conditional probability* of A by B, denoted by $P(A|B)$.

Suppose A and B are two events and $P(B) > 0$, the conditional probability $P(A|B)$ satisfies the following three requirements:

(1) Non-negative: For any event A, $P(A|B) \geqslant 0$ holds.

(2) Normality: For a certain event S, $P(S|B) = 1$ holds.

(3) Conformable additivity: Suppose $A_1, A_2, \ldots$ are mutually pairwise exclusive events, then $P(\bigcup_{i=1}^{\infty} A_i|B) = \sum_{i=1}^{\infty} P(A_i|B)$.

Besides these, $P(AB) = P(A)P(B|A)$ can be easily deducted from the definition of conditional probability. And this can be further extended to the case of more than two events, for example, $P(ABC) = P(C|AB)P(B|A)P(A)$.

Definition 13.7 (Joint Probability). The *joint probability* of two events A and B is the probability of the intersection of A and B. In other words, for two random variables X and Y, $P(X = a, Y = b)$ denotes the joint probability that $X = a$ and $Y = b$. Such a probability, that more than one event occurs and all these events occur simultaneously, is called a *joint probability*.

Definition 13.8 (Marginal Probability). The probability that an event A occurs alone, i.e., probabilities such as $P(X = a)$ or $P(Y = b)$, related to only one random variable, are called *marginal probabilities*.

Example 13.3   A company has 2,000 staff members, including 1,200 males and 800 females. Among them, 100 males and 50 females hold doctoral degrees. One staff from the company is randomly selected. Let A denote an event that the selected staff holds a doctoral degree, and let B denote an event that the selected staff is a female. Then we have

(1) $P(A) = \dfrac{100 + 50}{2000} = \dfrac{3}{40}$;

(2) $P(B) = \dfrac{800}{2000} = \dfrac{2}{5}$;

(3) AB is the event that the selected staff is a female with a doctoral degree,

so, $P(AB) = \dfrac{50}{2000} = \dfrac{1}{40}$;

(4) $A|B$ is the event that the selected staff holds a doctoral degree if the selected staff is a female. So, $P(A|B) = \dfrac{P(AB)}{P(B)} = \dfrac{1}{40} \times \dfrac{5}{2} = \dfrac{1}{16}$.

13.2.2   Total Probability Theorem and Bayes, Theorem

Definition 13.9 (Total Probability Theorem). Suppose a sample space Ω is the union of a series of mutually pairwise exclusive events A_1, A_2, ..., A_n. Then for any event B, $P(B) = \sum_{k=1}^{n} P(B \cap A_k)$. This theorem is called the *total probability theorem*.

Example 13.4   As shown in Figure 13.1a, the sample space Ω is the union of A and $\bar{A}$. As such, the event B can be split into two parts, as shown in Figure 13.1b, i.e., $P(B) = P(B \cap A) + P(B \cap \bar{A})$.

Definition 13.10 (Bayes, Theorem). $P(B|A) = P(A|B)P(B)/P(A)$.

It is very straight forward to proof Bayes, theorem, $P(B|A) = P(BA)/P(A) = P(A|B)P(B)/P(A)$.

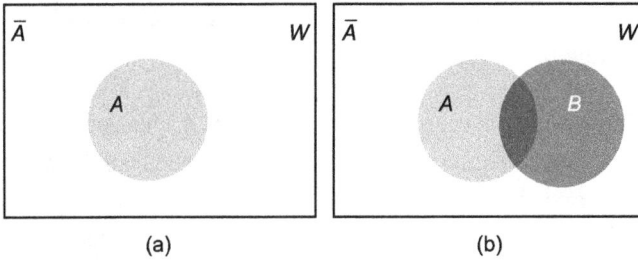

Figure 13.1 Total probability

Example 13.5 When conducting a blood test, the probability of being positive for a patient carrying the virus is 0.95, and the probability of being positive for a healthy person who does not carry the virus is 0.01. Now suppose the probability of a general person to have the virus is 0.3%, and find out the probability that a person does have the disease if the blood test result is positive.

Suppose $A = \{$positive$\}$ and $B = \{$carrying the virus$\}$, then $P(B) = 0.003$, $P(\overline{B}) = 0.997$, $P(A|B) = 0.95$, $P(A|\overline{B}) = 0.01$. According to the Bayes, theorem,

$$P(B|A) = \frac{P(B)P(A|B)}{P(B)P(A|B) + P(\overline{B})P(A|\overline{B})}$$

$$= \frac{0.003 \times 0.95}{0.003 \times 0.95 + 0.997 \times 0.01}$$

$$\approx 0.222$$

13.2.3 Independence

Definition 13.11 (Independent Event). Suppose A and B are two events of a random experiment E. If $P(AB) = P(A)P(B)$, then events A and B are mutually independent.[1]

---

[1] Independent events and mutually exclusive events are different. A and B are independent events, means that the probability that A occurs does not affect the occurrence of B. A and B are mutually exclusive events means that if A occurs, then B must not occur. The occurrence of A affects B. For example, sunny and rainy days are mutually exclusive events, while sunny days and eating lunch are independent events.

Corollary 13.1 If A and B are independent, then the following event pairs are also mutually independent: $\{\overline{A}, B\}$, $\{A, \overline{B}\}$, $\{\overline{A}, \overline{B}\}$.

Example 13.6 Suppose there are 5 white balls and 3 black balls in a box; take a ball from the box and put it back. Let A denote the event that "a black ball is taken at the first time", and B denote the event that "a white ball is taken at the second time". These two events are independent of each other, because $P(A) = \dfrac{3}{8}$, $P(B|A) = \dfrac{5}{8} = P(B)$, and we have $P(AB) = P(A)P(B|A) = \dfrac{3}{8} \times \dfrac{5}{8} = P(A)P(B)$.

13.3 Random Variable and Distribution Function

13.3.1 Fundamental Concepts

Definition 13.12 (Random Variable). Suppose E is a random experiment and Ω is its sample space. If for each sample $\omega \in \Omega$, there is a unique corresponding real number $X(\omega)$, then the real variable $X(\omega)$ is called a random variable denoted as X.

There are two types of random variables, discrete and continuous.

Definition 13.13 (Distribution Function). Let X be a random variable and x be an arbitrary real number, then the function $F(x) = P\{X \leqslant x\}$ is called the *distribution function* of X, also *cumulative distribution function*. The distribution function $F(x)$ of a random variable is a function defined by the probability of the event $\{X \leqslant x\}$. It is a common function with the value of the independent variable x ranging in $(-\infty, +\infty)$, and its value ranges in $[0, 1]$.

Distribution functions are very important in probability theory and statistics for description of the statistical laws of random variables.

Here are some common properties of a distribution function $F(x)$:

(1) $0 \leqslant F(x) \leqslant 1$ and $\lim_{x \to -\infty} F(x) = 0$, $\lim_{x \to +\infty} F(x) = 1$.

(2) $P\{x_1 < X \leqslant x_2\} = P\{X \leqslant x_2\} - P\{X \leqslant x_1\} = F(x_2) - F(x_1)$.

(3) $P\{X < b\} = P\left\{\lim_{n\to+\infty}\left(X \leqslant b - \dfrac{1}{n}\right)\right\} = \lim_{n\to+\infty} P\left\{X \leqslant b - \dfrac{1}{n}\right\}$

$\qquad = \lim_{n\to+\infty} F\left(b - \dfrac{1}{n}\right).$

13.3.2 Discrete Random Variables and Their Distribution Functions

Definition 13.14 (Discrete Random Variable). If a random variable X takes values from a finite or a countably infinite set of discrete values, X is called a discrete random variable.

Suppose all the potential values of a discrete random variable X are $x_k (k = 1, 2, \ldots)$, and the probability of the event $\{X = x_k\}$ is $P\{X = x_k\} = p_k (k = 1, 2, \ldots)$, in which $0 \leqslant p_k \leqslant 1$, and $\sum_{k=1}^{\infty} p_k = 1$, then $P\{X = x_k\} = p_k (k = 1, 2, \ldots)$ is called the probability density function (probability mass function, probability distribution) of the random variable X, and the distribution function of the discrete random variable X is

$$F(x) = P\{X \leqslant x\} = \sum_{x_k \leqslant x} P\{X = x_k\} = \begin{cases} 0 & x < x_1 \\ p_1 & x_1 \leqslant x < x_2 \\ \vdots & \vdots \\ \sum_{i=1}^{k} p_i & x_k \leqslant x < x_{k+1} \\ \vdots & \vdots \\ 1 & x \geqslant x_n \end{cases}$$

in which the sum of p_k is for all k that $x_k \leqslant x$.

Example 13.7 The probability distribution function of a random variable X is:

$$F(x) = \begin{cases} 0 & x < 0 \\ x/2 & 0 \leqslant x < 1 \\ 2/3 & 1 \leqslant x < 2 \\ 11/12 & 2 \leqslant x < 3 \\ 1 & 3 \leqslant x \end{cases}$$

we have

(1) $P\{X < 3\} = \lim_{n \to +\infty} P\left\{X \leqslant 3 - \dfrac{1}{n}\right\} = \lim_{n \to +\infty} F\left(3 - \dfrac{1}{n}\right) = \dfrac{11}{12}.$

(2) $P\{X = 1\} = P\{X \leqslant 1\} - P\{X < 1\} = F(1) - \lim_{n \to \infty} F\left(1 - \dfrac{1}{n}\right)$

$$= \dfrac{2}{3} - \dfrac{1}{2} = \dfrac{1}{6}.$$

(3) $P\left\{X > \dfrac{1}{2}\right\} = 1 - P\left\{X \leqslant \dfrac{1}{2}\right\} = 1 - F\left(\dfrac{1}{2}\right) = \dfrac{3}{4}.$

(4) $P\{2 < X \leqslant 4\} = F(4) - F(2) = \dfrac{1}{12}.$

Common discrete distributions include binomial distribution, polynomial distribution, Bernoulli distribution, Poisson distribution, etc.

13.3.3 Continuous Random Variables and Their Cumulative Distribution Functions

Definition 13.15 (Continuous Random Variable). If a random variable X takes an infinite and uncountable number of continuous values, X is called a continuous random variable.

Definition 13.16 (Probability Density Function). $F(x)$ is the distribution function of a continuous random variable X, if for any x, there exists $\varphi(x) \geqslant 0$, so that $F(x) = \displaystyle\int_{-\infty}^{x} \varphi(t)\,dt$, and $\displaystyle\int_{-\infty}^{+\infty} \varphi(t)\,dt = 1$, then $\varphi(x)$ is called the probability density function of X.

Common continuous distributions include uniform distribution, normal distribution, exponential distribution, χ^2 distribution, F distribution, etc.

Taking normal distribution as an example, a random variable X is a normal random variable with parameters μ and σ^2, denoted by $X \sim N(\mu, \sigma^2)$, if the probability density function of X is

$$f(x) = \dfrac{1}{\sqrt{2\pi}\sigma} e^{-(x-\mu)^2/2\sigma^2}, \quad -\infty < x < +\infty$$

This density function is a bell-shaped curve symmetrical to μ.

The normal distribution with $\mu = 0$ and $\sigma^2 = 1$ is called standard normal distribution. Its density function is

$$f(x) = \frac{1}{\sqrt{2\pi}} e^{-x^2/2}, -\infty < x < +\infty$$

and its distribution function is denoted by $\Phi(x)$, i.e.,

$$\Phi(x) = \frac{1}{\sqrt{2\pi}} \int_{-\infty}^{x} e^{-t^2/2} dt, -\infty < x < +\infty$$

For any non-negative real number x, $\Phi(x)$ can be obtained by looking up the standard normal distribution table. For a negative x, $\Phi(x) = 1 - \Phi(-x)$.

Corollary 13.2   For a normal distribution random variable $X \sim N(\mu, \sigma^2)$, $Z = (X - \mu)/\sigma$ follows a standard normal distribution, and the distribution function of X can be written as

$$F_X(a) = P\{X \leqslant a\} = P\left\{\frac{X - \mu}{\sigma} \leqslant \frac{a - \mu}{\sigma}\right\} = \Phi\left(\frac{a - \mu}{\sigma}\right)$$

Example 13.8 X follows a normal distribution with parameters $\mu = 3$ and $\sigma^2 = 9$, then we have

(1) $P\{2 < X < 5\} = P\left\{\frac{2-3}{3} < \frac{X-3}{3} < \frac{5-3}{3}\right\} = P\left\{-\frac{1}{3} < Z < \frac{2}{3}\right\}$

$$= \Phi\left(\frac{2}{3}\right) - \Phi\left(-\frac{1}{3}\right) = \Phi\left(\frac{2}{3}\right) - \left[1 - \Phi\left(\frac{1}{3}\right)\right] \approx 0.3779.$$

(2) $P\{X > 0\} = P\left\{\frac{X-3}{3} > \frac{0-3}{3}\right\} = P\{Z > -1\} = 1 - \Phi(-1)$

$$= \Phi(1) \approx 0.8413.$$

(3) $P\{|X - 3| > 6\} = P\{X > 9\} + P\{X < -3\}$

$$= P\left\{\frac{X-3}{3} > \frac{9-3}{3}\right\} + P\left\{\frac{X-3}{3} < \frac{-3-3}{3}\right\}$$

$$= P\{Z > 2\} + P\{Z < -2\}$$

$$= 1 - \Phi(2) + \Phi(-2)$$

$$= 2[1 - \Phi(2)] \approx 0.0456.$$

13.4   Mathematical Characteristics

13.4.1   Mathematical Expectation

Definition 13.17 (Mathematical Expectation of A Discrete Random Variable). Mathematical expectation of a discrete random variable X is defined as:

$$E[X] = \sum_x xp(x)$$

Where $p(x)$ is the probability distribution of X.

Corollary 13.3   If Y is an image of a discrete random variable X under the function, i.e., $Y = g(X)$, the mathematical expectation of Y is

$$E[Y] = \sum_x g(x)p(x)$$

Definition 13.18 (Mathematical Expectation of A Continuous Random Variable). Mathematical expectation of a continuous random variable X is defined as:

$$E[X] = \int_{-\infty}^{+\infty} xf(x)dx$$

where $f(x)$ is the probability density function of X.

Corollary 13.4   If Y is a function of a continuous random variable X, i.e., $Y = g(X)$, the mathematical expectation of Y is,

$$E[Y] = \int_{-\infty}^{+\infty} g(x)f(x)dx$$

The mathematical expectation represents the average level of the values of a random variable.

The above definitions and corollaries can be extended to multi-dimensional spaces. Taking a two-dimensional space as an example. X and Y are two random variables, then (X, Y) is a two-dimensional random vector.

If the two-dimensional probability distribution of a two-dimensional discrete random vector (X, Y) is $p(x, y)$, and g is a binary function, then

$$E[g(X, Y)] = \sum_y \sum_x g(x, y)p(x, y)$$

If the joint probability density function of a two-dimensional continuous random vector (X, Y) is $f(x, y)$, and g is a binary function, then

$$E[g(X,Y)] = \int_{-\infty}^{+\infty} \int_{-\infty}^{+\infty} g(x, y) f(x, y) dxdy$$

Example 13.9 Suppose two arbitrary points X and Y are independently selected from a line segment with length S, then X and Y are independent and uniformly distributed on the line segment with the joint probability density function of (X, Y) as $f(x, y) = \frac{1}{S^2}, 0 < x < S, 0 < y < S$. The expectation of the distance of X and Y is

$$E(|X - Y|) = \frac{1}{S^2} \int_0^S \int_0^S |x - y| dydx$$
$$= \frac{1}{S^2} \int_0^S \int_0^x (x - y) dydx + \frac{1}{S^2} \int_0^S \int_x^S (y - x) dydx$$
$$= \frac{S}{3}$$

13.4.2 Variance and Correlation Coefficient

Definition 13.19 (Variance of A Discrete Random Variable). The variance of a discrete random variable X is defined as:

$$\text{Var}(X) = E[X - E(X)]^2$$

Definition 13.20 (Variance of A Continuous Random Variable). The variance of a continuous random variable X is defined as:

$$\text{Var}(X) = \int_{-\infty}^{+\infty} [x - E(X)]^2 f(x) dx$$

where $f(x)$ is the probability density function of X.

The variance of a random variable is a positive number. If the possible values of a random variable X are near its expectation, the variance of X is small. Otherwise, it is large. Variance represents the degree of dispersion of the values of a random variable.

Example 13.10 The numbers of rings shot by two shooters A and B are X and Y. The probability distributions of X and Y are shown as Table 13.1.

Table 13.1 The numbers of rings of A and B

X	10	9	8	7	6	5	4	3	2	1	0
$P(X)$	0.3	0.2	0.1	0.1	0.05	0.05	0.04	0.03	0.02	0.01	0.1
Y	10	9	8	7	6	5	4	3	2	1	0
$P(Y)$	0.5	0.3	0.06	0.03	0.02	0.02	0.02	0.02	0.01	0.01	0.01

We can calculate $\mathrm{Var}(X)$ and $\mathrm{Var}(Y)$ by:

$$
\begin{aligned}
E(X) &= 0.3 \times 10 + 0.2 \times 9 + 0.1 \times 8 + 0.1 \times 7 + 0.05 \times 6 + 0.05 \times 5 \\
&\quad + 0.04 \times 4 + 0.03 \times 3 + 0.02 \times 2 + 0.01 \times 1 + 0.1 \times 0 \\
&= 7.15 \\
E(Y) &= 0.5 \times 10 + 0.3 \times 9 + 0.06 \times 8 + 0.03 \times 7 + 0.02 \times 6 + 0.02 \times 5 \\
&\quad + 0.02 \times 4 + 0.02 \times 3 + 0.01 \times 2 + 0.01 \times 1 + 0.01 \times 0 \\
&= 8.78 \\
\mathrm{Var}(X) &= 0.3 \times (10 - 7.15)^2 + 0.2 \times (9 - 7.15)^2 + 0.1 \times (8 - 7.15)^2 \\
&\quad + 0.1 \times (7 - 7.15)^2 + 0.05 \times (6 - 7.15)^2 + 0.05 \times (5 - 7.15)^2 \\
&\quad + 0.04 \times (4 - 7.15)^2 + 0.03 \times (3 - 7.15)^2 + 0.02 \times (2 - 7.15)^2 \\
&\quad + 0.01 \times (1 - 7.15)^2 + 0.1 \times (0 - 7.15)^2 \\
&= 10.4275 \\
\mathrm{Var}(Y) &= 0.5 \times (10 - 8.78)^2 + 0.3 \times (9 - 8.78)^2 + 0.06 \times (8 - 8.78)^2 \\
&\quad + 0.03 \times (7 - 8.78)^2 + 0.02 \times (6 - 8.78)^2 + 0.02 \times (5 - 8.78)^2 \\
&\quad + 0.02 \times (4 - 8.78)^2 + 0.02 \times (3 - 8.78)^2 + 0.01 \times (2 - 8.78)^2 \\
&\quad + 0.01 \times (1 - 8.78)^2 + 0.01 \times (0 - 8.78)^2 \\
&= 4.2916
\end{aligned}
$$

Expectation and variance describe mathematical characteristics of a single random variable. Covariance describes the correlation of two random variables.

Definition 13.21 (Covariance). The covariance of two random variables X and Y, $\mathrm{Cov}(X, Y)$, is defined as

$$
\mathrm{Cov}(X, Y) = E[(X - E(X))(Y - E(Y))]
$$

Covariance can be negative, zero or positive. If two random variables are independent, the covariance of the two random variables are zero. Some other properties of covariance are given below. Readers can prove them as exercises.

(1) $\mathrm{Cov}(X, Y) = \mathrm{Cov}(Y, X)$;
(2) $\mathrm{Cov}(X, X) = \mathrm{Var}(X)$;
(3) $\mathrm{Cov}(aX, Y) = a\,\mathrm{Cov}(X, Y)$;
(4) $\mathrm{Cov}(\sum_{i=1}^{n} X_i, \sum_{j=1}^{m} Y_j) = \sum_{i=1}^{n} \sum_{j=1}^{m} \mathrm{Cov}(X_i, Y_j)$.

Definition 13.22 (Correlation Coefficient). Suppose X and Y are two random variables, both with positive variances, then the correlation coefficient of X and Y is defined as

$$\rho(X, Y) = \frac{\mathrm{Cov}(X, Y)}{\sqrt{\mathrm{Var}(X)\,\mathrm{Var}(Y)}}$$

Correlation coefficient is normalized covariance.

13.5 Applications in GIS

There are many examples of probability theory applied in GIS. The most systematic achievements are the theory and methodologies of spatial statistics, in which mathematical statistics supported by probability theory are systematically combined with spatial analysis methodologies. Some typical examples are: the mathematical expectation of multiple measures is applied to estimate the real geographic position; the variances of two surveying methods are applied to compare their stability; kernel density function is applied to make heat maps; correlation coefficient is applied to describe and calculate spatial correlation, etc.

13.6 Exercises

Exercise 13.1 Suppose two events A and B are independent, and $P((\overline{A})\overline{B}) = \frac{1}{9}$, $P(A\overline{B}) = P((\overline{A})B)$, find $P(A)$ and $P(B)$.

Exercise 13.2 There are four cards, the first one has red dots only, the second one has yellow dots only, the third one has green dots only, and the fourth one has red, yellow and green dots. A, B, and C represent the three events that a card randomly drawn has red, yellow or green dots respectively. Investigate the independence of A, B, and C.

Exercise 13.3 Suppose $X_1, X_2, ..., X_n$ are n random variables. Prove that

$$\text{Var}\left(\sum_{i=1}^{n} X_i\right) = \sum_{i=1}^{n} \text{Var}(X_i) + 2 \sum_{1 \leqslant i < j \leqslant n} \text{Cov}(X_i, X_j)$$

and if $X_1, X_2, ..., X_n$ are mutually independent, then, $P(\sum_{i=1}^{n} X_i) = \sum_{i=1}^{n} PX_i$.

Exercise 13.4 Prove that for any two events A and B, $P(B - A) = P(B) - P(AB)$ holds.

Chapter 14

Statistical Discriminant Analysis

In industrial manufacturing, scientific research, and commercial activities, it is often necessary to choose an optimized decision among several options. For example, in remote sensing images, different land cover types are imaged as pixels of different colors. Knowing the land cover type of some pixels, a certain rule can be set up so that the pixels of the entire image can be categorized into cultivated land, forest land, grassland, residential land, roads, etc. The results can serve governmental departments for management and decision-making purposes. This is a common example to make a categorization decision. The basic process is to establish a categorization decision model based on some known instances to determine the categorization of unknown instances. This process model is called discriminant analysis in mathematical statistics.

Starting from the basic concept of statistical discriminant analysis, this chapter introduces some common methods with some examples to enhance readers' understanding.

14.1   Basic Concepts

Definition 14.1 (Statistical Discriminant Analysis). Based   on   instance data of known categories, a statistically optimized categorization rule is established to determine the categories of new instances. The process of establishing these statistically optimized rules is called statistical discriminant analysis.

Example 14.1   Based on locally collected historical weather data, predicting whether tomorrow will be cloudy, sunny or rainy, is a matter of statistical discriminant question.

Statistical discriminant analysis is a statistical solution to a type of classification problem. In the field of machine learning, there are two basic models of classification, i.e., supervised classification and unsupervised classification. The supervised classification is to learn through the existing data and obtain an optimized or optimal rule according to an objective function. This rule maps each example (existing or unknown) to a predefined class. Supervised classification and statistical discriminant analysis both solve the same type of classification problems.

Like machine learning, many other fields also study classification problems, and some even form their own terminology. Sometimes these terms, even the same term, are slightly different in different fields.

As a statistical classification solution, the biggest difference from solutions of other fields, is that statistical discriminant is based on statistically significant optimization goals, for example, the largest probability of correct categorization, the lowest probability of incorrect categorization, or the smallest expected loss of incorrect categorization, etc. In many cases, it is difficult to have statistically strict independent and identical distributions, or even to give a probability distribution.

Therefore, in practical applications, statistical discriminant analysis is often simplified and directly optimized using a simple objective function without special consideration of probability distribution.

In statistics, there is another process, which is also related to classification. Given a series of samples, the process of aggregating these samples into several categories through statistical optimization rules is called statistical clustering.

Example 14.2   Based on the analysis of meteorological data over the past 365 days, what types of local weather can be determined? For example, according to the daily maximum temperature, average temperature or minimum temperature, it may be divided into hot and cold days; according to the record of sunlight and rainfall, it may be divided into sunny, cloudy, light rain, heavy rain and storm, etc.

Therefore, the difference between statistical discriminant analysis and statistical clustering is obvious. Statistical discriminant analysis is to build classification rules from known instances. Generally, how many and which categories are known and the goal is to classify new samples. Statistical clustering is to aggregate known instances into several categories according to certain rules. It is not known in advance which categories and how many categories to aggregate.

Supervised classification in the field of machine learning corresponds to statistical classification; unsupervised classification is a process corresponding to statistical clustering.

14.2 Evaluation of A Statistical Discriminant Model

Evaluating a statistical discriminant model is generally judged by the probability of correct or incorrect decisions, or the expected value of a loss function. However, sometimes it is difficult to determine the exact distribution of variables, and it is not possible to accurately calculate the probability of a correct or wrong decision or the expected value of a loss. For this reason, the categorized results of the known instances are compared with the true value, which is also used in practice to objectively evaluate the quality of a discriminant model. This is consistent with the evaluation of classification methods in the field of machine learning.

Some objective evaluation indicators are accuracy, precision, recall, F-value, etc.

Definition 14.2 (True Positive, TP). A TRUE instance is categorized positively (correctly).

Definition 14.3 (True Negative, TN). A FALSE instance is categorized negatively (wrongly).

Definition 14.4 (False Positive, FP). A FALSE instance is categorized positively (type I error).

Definition 14.5 (False Negative, FN). A TRUE instance is categorized negatively (type II error).

Definition 14.6 (Confusion Matrix). The above four indicators organized into a matrix, is called a *confusion matrix*.

Definition 14.7 (Accuracy). $\text{ACCURACY} = \dfrac{\text{TP} + \text{TN}}{\text{TP} + \text{TN} + \text{FP} + \text{FN}}.$

Definition 14.8 (Precision). $\text{PRECISION} = \dfrac{\text{TP}}{\text{TP} + \text{FP}}.$

Definition 14.9 (Recall). $\text{RECALL} = \dfrac{\text{TP}}{\text{TP} + \text{FN}}.$

14.3   Distance-Based Discriminant Analysis

The basic idea of distance-based discriminant analysis is to calculate the distances between the instance to be discriminated and the known categories and find the closest category, and make a judgement that the sample to be discriminated belongs to this category.

The core of the distance-based discriminant mode is the definition of distance (see Definition 9.1). Any distance that fulfills the following three properties can act as the basis for discriminant model.

(1) Non-negative: $d(x, y) \geqslant 0$;

(2) Symmetry: $d(x, y) = d(y, x)$;

(3) Triangular inequality: $d(x, y) + d(y, z) \geqslant d(x, z)$, for any x, y, and z.

The most frequently used distance is the Euclidean distance.

$$D = \sqrt[2]{(x_1 - x_2)^2 + (y_1 - y_2)^2}$$

The distance of a point to a category can be calculated by the distance between the point and the centroid, gravity center, the nearest point, the most far point of the category, etc.

Sometimes, distance is also used to develop other rules. For example, for a point x to be discriminated, under the Euclidean distance, the nearest k points to x are picked out. Then the k points picked out are counted by which category they come from. The category with the biggest number of the k picked-out points, is judged as the category of x. This algorithm is in many other fields called k-Nearest Neighbor (kNN). The procedure of kNN is illustrated by Figure 14.1.

Example 14.3   Supposing by known instances, the centroid of first category is $x_1 = (0.1, 2.3)$, the centroid of the second category is $x_2 = (2.8, 2.5)$, and the centroid of the third category is $x_3 = (1.5, 0.3)$. By using the Euclidean distance, determine to which category a new instance $x = (1.2, 1.0)$ belongs to?

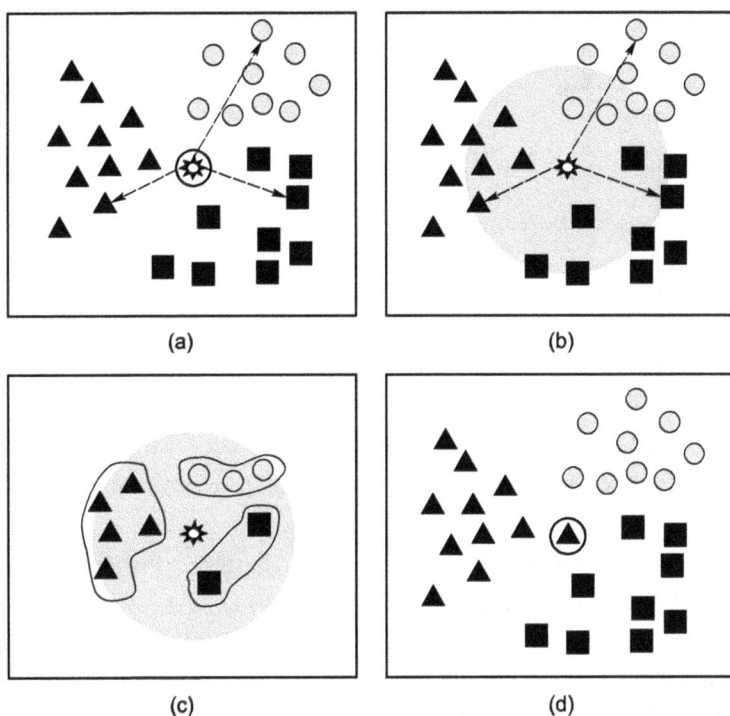

Figure 14.1 kNN algorithm

Solution:

$$d(x, x_1) = \text{sqrt}((1.2 - 0.1)^{\wedge}2 + (1.0 - 2.3)^{\wedge}2) \approx 1.703,$$
$$d(x, x_2) = \text{sqrt}((1.2 - 2.8)^{\wedge}2 + (1.0 - 2.5)^{\wedge}2) \approx 2.193,$$
$$d(x, x_3) = \text{sqrt}((1.2 - 1.5)^{\wedge}2 + (1.0 - 0.3)^{\wedge}2) \approx 0.762.$$

So, the judgement is that x belongs to the third category.

14.4 Fisher Discriminant Analysis

The basic idea of Fisher discriminant analysis is to project the points in the multi-dimensional space to a one-dimensional space, and find a threshold to split the one-dimensional space, so that the variance between the classes is the largest, and the variances within the classes are the smallest, as shown in Figure 14.2. The projection from the multi-dimensional space to

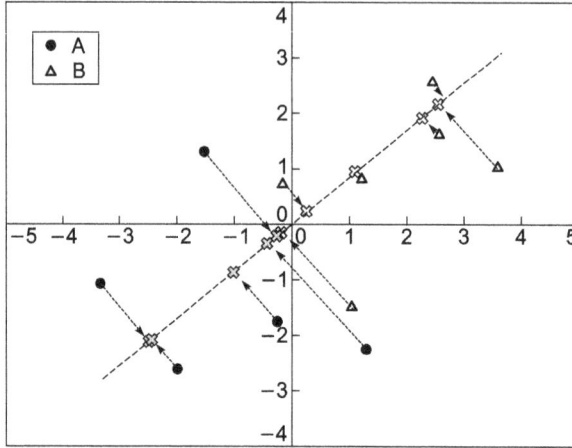

Figure 14.2　The projection from the two-dimensional space to a one-dimensional space

a one-dimensional space is a linear function. Therefore, Fisher discriminant analysis is also called linear discriminant analysis (LDA).

Let's take an example of a two-category classification problem, to illustrate the principle of the Fisher discriminant analysis. Suppose there are two categories of k-dimensional instance data sets, with N_0 and N_1 respectively, denoted by $X_0 = \{\boldsymbol{x}_{0i}, i = 1, 2, \ldots, N_0\}$ and $X_1 = \{\boldsymbol{x}_{1i}, i = 1, 2, \ldots, N_1\}$. Their means and variances are denoted by $\boldsymbol{\mu}_0, \boldsymbol{\mu}_1$ and $\boldsymbol{\Sigma}_0, \boldsymbol{\Sigma}_1$ respectively.

$$\boldsymbol{\Sigma}_j = \sum_{\boldsymbol{x} \in X_j} (\boldsymbol{x} - \boldsymbol{\mu}_j)(\boldsymbol{x} - \boldsymbol{\mu}_j)^{\mathrm{T}}, \quad \boldsymbol{\mu}_j = \frac{1}{N_j} \sum_{\boldsymbol{x} \in X_j} \boldsymbol{x}, j \in \{0, 1\}$$

Suppose all the samples are projected to a line $\boldsymbol{w}$, i.e., for any sample $\boldsymbol{x}$, the projection is $\boldsymbol{w}^{\mathrm{T}}\boldsymbol{x}$. By doing so, the centroid points of the two categories $\boldsymbol{\mu}_0$ and $\boldsymbol{\mu}_1$ are projected to $\boldsymbol{w}^{\mathrm{T}}\boldsymbol{\mu}_0$ and $\boldsymbol{w}^{\mathrm{T}}\boldsymbol{\mu}_1$, and the variances of the projected points of the two categories are $\boldsymbol{w}^{\mathrm{T}}\boldsymbol{\Sigma}_0\boldsymbol{w}$ and $\boldsymbol{w}^{\mathrm{T}}\boldsymbol{\Sigma}_1\boldsymbol{w}$.

The goal of the Fisher discriminant analysis is to disperse the two categories as far as possible, i.e., to maximize $\|\boldsymbol{w}^{\mathrm{T}}\boldsymbol{\mu}_0 - \boldsymbol{w}^{\mathrm{T}}\boldsymbol{\mu}_1\|^2 = \boldsymbol{w}^{\mathrm{T}}(\boldsymbol{\mu}_0 - \boldsymbol{\mu}_1)(\boldsymbol{\mu}_0 - \boldsymbol{\mu}_1)^{\mathrm{T}}\boldsymbol{w}$, and to aggregate the projected points within a category as close as possible, i.e., to minimize the sum of the variances of the projected points within the two categories $\boldsymbol{w}^{\mathrm{T}}\boldsymbol{\Sigma}_0\boldsymbol{w} + \boldsymbol{w}^{\mathrm{T}}\boldsymbol{\Sigma}_1\boldsymbol{w} = \boldsymbol{w}^{\mathrm{T}}(\boldsymbol{\Sigma}_0 + \boldsymbol{\Sigma}_1)\boldsymbol{w}$. In general, the goal is to maximize the following function:

$$\frac{\boldsymbol{w}^{\mathrm{T}}(\boldsymbol{\mu}_0 - \boldsymbol{\mu}_1)(\boldsymbol{\mu}_0 - \boldsymbol{\mu}_1)^{\mathrm{T}}\boldsymbol{w}}{\boldsymbol{w}^{\mathrm{T}}(\boldsymbol{\Sigma}_0 + \boldsymbol{\Sigma}_1)\boldsymbol{w}}$$

Let's denote the disperse matrix within categories as $S_w = \Sigma_0 + \Sigma_1$, and the disperse matrix between the two categories by $S_b = (\mu_0 - \mu_1)(\mu_0 - \mu_1)^T$, then the above goal function can be rewritten as:

$$\frac{w^T S_b w}{w^T S_w w}$$

According to the Rény entropy formula, the maximum value of the above expression is the biggest eigenvalue of the matrix $S_w^{-\frac{1}{2}} S_b S_w^{-\frac{1}{2}}$, when w is the corresponding eigenvector, i.e., $S_w^{-1}(\mu_0 - \mu_1)$. For a new sample x, to determine which category that x belongs to, all the necessary is to calculate the distance between the projected point $w^T x$ of x on w, and the projected points $w^T \mu_0$ and $w^T \mu_1$ of the two categories' centroid points on w. x belongs to the first category X_0, if $(w^T x - w^T \mu_0) < (w^T x - w^T \mu_1)$; otherwise x belongs to the second category X_1.

Example 14.4 As shown in Figure 14.2, in a two-dimensional space, the first category A has 5 instances (dots), i.e., $(-3.38, -0.61)$, $(-5.53, -1.07)$, $(-4.13, -2.30)$, $(-3.26, -0.76)$ and $(-4.68, -0.25)$, and the second category B has 6 instances (triangles), i.e., $(4.58, -0.06)$, $(2.86, 1.64)$, $(3.21, -0.84)$, $(5.50, -1.25)$, $(3.94, -0.91)$ and $(5.55, 2.29)$. Give the Fisher discriminant function and find which category the two new points $x_1(-3.54, -2.06)$ and $x_2(5.04, -1.12)$ belong to.

Solution: $N_0 = 5$, $N_1 = 6$; we can calculate μ_0 and μ_1, Σ_0 and Σ_1:

$$\mu_0 = (-4.19, -1.00)$$
$$\mu_1 = (4.27, 0.15)$$
$$\Sigma_0 = \begin{pmatrix} 6.48 & 0.27 \\ 0.27 & 10.90 \end{pmatrix}$$
$$\Sigma_1 = \begin{pmatrix} 3.57 & 0.20 \\ 0.2 & 2.47 \end{pmatrix}$$

so,

$$S_b = (\mu_0 - \mu_1)(\mu_0 - \mu_1)^T = 73.05$$
$$S_w = \Sigma_0 + \Sigma_1 = \begin{pmatrix} 10.05 & 0.47 \\ 0.47 & 13.37 \end{pmatrix}$$

We can easily get $\boldsymbol{w}$:

$$\boldsymbol{w} = \mathbf{S}_w^{-1}(\boldsymbol{\mu}_0 - \boldsymbol{\mu}_1) = (0.84, 0.06)$$

Then we can get:

$$\boldsymbol{w}^{\mathrm{T}}\boldsymbol{\mu}_0 = 3.58$$
$$\boldsymbol{w}^{\mathrm{T}}\boldsymbol{\mu}_1 = 3.60$$
$$\boldsymbol{w}^{\mathrm{T}}\boldsymbol{x}_1 = -3.09$$
$$\boldsymbol{w}^{\mathrm{T}}\boldsymbol{x}_2 = 4.17$$

For $\boldsymbol{x}_1$, we have

$$d_0 - d_1 = |\boldsymbol{w}^{\mathrm{T}}\boldsymbol{x}_1 - \boldsymbol{w}^{\mathrm{T}}\boldsymbol{\mu}_0| - |\boldsymbol{w}^{\mathrm{T}}\boldsymbol{x}_1 - \boldsymbol{w}^{\mathrm{T}}\boldsymbol{\mu}_1| < 0$$

Since $d_0 < d_1$, $\boldsymbol{x}_1$ belongs to the first category A.

For $\boldsymbol{x}_2$, we have

$$d_0 - d_1 = |\boldsymbol{w}^{\mathrm{T}}\boldsymbol{x}_2 - \boldsymbol{w}^{\mathrm{T}}\boldsymbol{\mu}_0| - |\boldsymbol{w}^{\mathrm{T}}\boldsymbol{x}_2 - \boldsymbol{w}^{\mathrm{T}}\boldsymbol{\mu}_1| > 0$$

Since $d_0 > d_2$, $\boldsymbol{x}_2$ belongs to the second category B.

14.5   Logistic Regression Discriminant Model

In the two-class discrimination problem, a regression model can be established by assigning the value of one class to 0 and the other class to 1, like this: $y = f(x) + \varepsilon$.

When we use a sigmoid S-type function $y = \dfrac{1}{1 + e^{-x}}$ as regression model as shown in Figure 14.3, the discriminant model is called logistic regression discriminant method.

To do that, we plug the sample into the following formula:

$$y = \frac{1}{1 + e^{-(\boldsymbol{w}^{\mathrm{T}}\boldsymbol{x}+b)}}$$

To solve the maximum likelihood estimate $\hat{\boldsymbol{w}}$ of this $\boldsymbol{w}$, a log-loss loss function is generally adopted, that is, to minimize the following formula:

$$L = \sum L_i, L_i = \begin{cases} -\log(\hat{y}_i), & y = 0 \\ -\log(1 - \hat{y}_i), & y = 1 \end{cases}, \quad i = 1, 2, \ldots, n$$

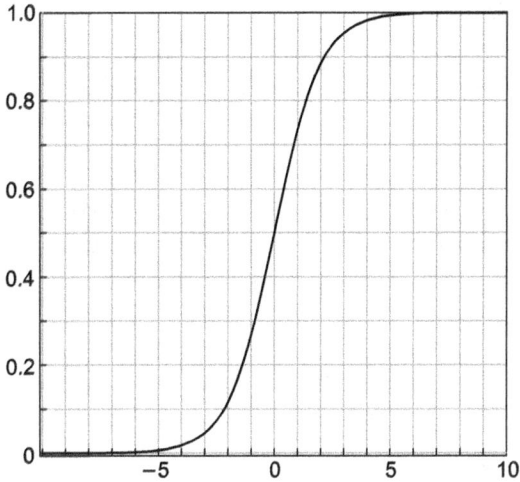

Figure 14.3 Sigmoid S-type function

Generally, the gradient descent method or Newton method can be used to calculate $\hat{\boldsymbol{w}}$ and $\hat{b}$.

The value range of a sigmoid S-type prediction function is $(0, 1)$. Generally, the default threshold is set to 0.5. If the regression value is greater than 0.5, the sample is considered to belong to category 1; otherwise, the sample is considered to belong to category 0.

Logistic regression discrimination is superior to large-scale data. Regardless of the relationship between variables, it can directly put any variable into the model for training. But the disadvantage is that it is easy to overfit and the classification accuracy is not too high. Common solutions include reducing variables (such as manually selecting some variables to apply) and normalization.

14.6 Bayes Discriminant Analysis

The basic procedure of Bayes discriminant analysis is, based on known prior probabilities, calculating the posterior probability of each category on the case of the new sample's occurrence. The maximum posterior probability of a category means that the new sample belongs to this category.

According to Bayes formula,

$$P(y_i|x) = \frac{P(x|y_i)P(y_i|x)}{P(x)} = \frac{P(x|y_i)P(y_i)}{\sum_i P(x|y_i)P(y_i)}$$

the inputs are the general probability of the occurrence of each category $P(y_i)$, i.e., prior probability, and the probability that x occurs with a category y_i, $P(x|y_i)$. The outputs are the probability that x comes from the category y_i, $P(y_i|x)$, i.e., posterior probability.

Example 14.5   For an area with frequent storms, ω_1 denotes a day that storms occur, and ω_2 denotes a day without storms. Based on long-term observations, the probability of a day with storms is $P(\omega_1) = 0.2$, and the probability of a day without storms is $P(\omega_2) = 1 - 0.2 = 0.8$. How to forecast tomorrow's storms in this area? Obviously, since $P(\omega_1) < P(\omega_2)$, no storm is a safe forecast if there is no additional information.

However, the occurrence of storms has much to do with the weather indicators, such as air pressure, moisture, temperature, etc., of the previous day. If the weather is abnormal in the previous day, storms are more likely to occur today. Let x denote the weather of the previous day, which takes value from {normal, abnormal}. Suppose we have the following statistics:

- The probability of storms with an abnormal weather in the previous day is 0.6, i.e., $P(x = \text{abnormal}|\omega_1) = 0.6$.
- The probability of storms with a normal weather in the previous day is 0.4, i.e., $P(x = \text{normal}|\omega_1) = 0.4$.
- The probability of no storm with an abnormal weather in the previous day is 0.1, i.e., $P(x = \text{abnormal}|\omega_2) = 0.1$.
- The probability of no storm with a normal weather in the previous day is 0.9, i.e., $P(x = \text{normal}|\omega_2) = 0.9$.

Suppose the weather is abnormal today, tell whether storms will happen tomorrow. The Bayes solution to this problem is to figure out the posterior probability $P(\omega_1|x = \text{abnormal})$. According to Bayes formula

$$P(\omega_1|x = \text{abnormal}) = \frac{P(x = \text{abnormal}|\omega_1)P(\omega_1)}{P(x = \text{abnormal})}$$

$$= \frac{P(x = \text{abnormal}|\omega_1)P(\omega_1)}{P(x = \text{abnormal}|\omega_1)P(\omega_1) + P(x = \text{abnormal}|\omega_2)P(\omega_2)}$$

$$= \frac{0.6 \times 0.2}{0.6 \times 0.2 + 0.1 \times 0.8} = 0.6$$

So, if the threshold is set to 0.5, the result should be "storms will happen".

The above Bayes discriminant approach is called naive Bayes classification in the field of machine learning, since this approach simply assumes that the variables involved are mutually conditional independent.

Sometimes, the risks caused by type I errors and type II errors cannot be considered equally. By introducing a loss function, making a decision by selecting the category with the least loss expectation, is called Bayes minimum risk discriminant.

Example 14.6 With the same background of Example 14.5, suppose that the cost to do preparation for the second day of possible storms is 16, while the loss of a storm day without any preparation in the previous day is 10. No cost or no loss occurs in other cases. If it is an abnormal day today, how to forecast tomorrow's storm to minimize the loss expectation?

Following Example 14.5, $P(\omega_1|x = \text{abnormal}) = 0.6$ and $P(\omega_2|x = \text{abnormal}) = 0.4$. If storms are forecasted, the expectation loss is $16 \times 0.4 = 6.4$. If no storm is forecasted, the expectation loss is $10 \times 0.6 = 6.0$.

The decision is that, although there will possibly be storms and cause loss, it costs more to do preparation. In general, the expected loss without preparation is less than the expected cost with preparation. So, this problem is somewhat not to forecast tomorrow's storm, but to make a wise decision to do preparation or not, for a generally minimum cost or loss.

14.7 Applications in GIS

The most common applications of statistical discriminant analysis in GIS is remote sensing image classification. The purpose of remote sensing image classification is to interpret the pixels in the image into different feature categories. Software currently available for image classification includes ENVI, ERDAS, and ArcGIS.

Obviously, statistical discriminant analysis can be widely used for decision analysis in GIS. For example, how to judge whether an area is flat, hilly, or mountainous? How to plan an area according to the suitability of plant species based on soil sampling data? How to divide a city into a grid of moderate size according to the information of the city's population, traffic, business, etc., and carry out urban management over these divided areas?

14.8   Exercises

Exercise 14.1   Suppose a discriminant model sets the threshold to 0.5 with a positive decision if the discriminant function outputs a value equal to or greater than 0.5. If the threshold is increased to 0.6, how will the precision and recall change?

Exercise 14.2   Suppose one more known instance $(3.21, 2.58)$ is given in Example 14.4, please give the new Fisher rule. And now, which category the new samples $x_1(-3.54, -2.06)$ and $x_2(5.04, -1.12)$ belong to?

Exercise 14.3   Give an example to illustrate the process of logistic regression discriminant model.

Exercise 14.4   Following Example 14.6, suppose the loss without preparation is 12, what should be the new rule?   And how about the loss is 18?

www.ingramcontent.com/pod-product-compliance
Lightning Source LLC
Chambersburg PA
CBHW081538190326
41458CB00015B/5585